POST-ANALYTIC *TRACTATUS*

Post-Analytic Tractatus establishes Wittgenstein's early work in the *Tractatus Logico-Philosophicus* as an invaluable source for exploring current debate on analytic philosophy in its origins, history, limits and relations with European philosophy. Drawing together new work from the leading figures in interpretation of the *Tractatus* – Conant and Diamond – with work by respected Wittgenstein commentators such as Kremer and Hutto, together with a reprint of a relevant and striking text by Brouwer, this timely collection offers a valuable resource for exploring the *Tractatus*' connections to approaches other than logical positivism, mathematical logic and formal semantics. Examining links with the work of Leibniz, Kant, Kierkegaard, Schopenhauer, Frege, Russell, James, Heidegger and others, the contributors consider key themes in twentieth-century philosophy including symbols and expression, language and metaphysics, objects and signs, logical form, structure and syntax, limits of philosophical discourse, Idealism and transcendental arguments, distinguishing sense and nonsense, showing and saying in communication, mysticism and transcendence in experience, ethical and aesthetic value, the worlds of solipsism and religion, philosophy as an activity and as a system.

Particularly timely in establishing the *Tractatus* as a source for comparable debates across Continental and Analytic philosophy, this collection will prove of value to scholars of twentieth-century philosophy, Wittgenstein, and Post-Analytic philosophy.

Post-Analytic *Tractatus*

Edited by
BARRY STOCKER

ASHGATE

Published by
Ashgate Publishing Limited
Gower House
Croft Road
Aldershot
Hants GU11 3HR
England

Ashgate Publishing Company
Suite 20
101 Cherry Street
Burlington, VT 05401-4405
USA

Ashgate website: http://www.ashgate.com

British Library Cataloguing in Publication Data
Post-analytic Tractatus
1. Wittgenstein, Ludwig, 1889-1951. Tractatus
Logico-philosophicus
I. Stocker, Barry
192

Library of Congress Cataloging-in-Publication Data
Post-analytic Tractatus / edited by Barry Stocker.
p. cm.
Includes bibliographical references and index.
ISBN 0-7546-1297-X
1. Wittgenstein, Ludwig, 1889-1951. Tractatus logico-philosophicus. 2. Logic, Symbolic and mathematical. 3. Language and languages--Philosophy. 4. Metaphysics. I. Stocker, Barry, 1966-

B3376.W563 T7366 2002
192--dc21

2002021442

ISBN 0 7546 1297 X

Printed and bound in Great Britain by MPG Books Ltd, Bodmin, Cornwall

Contents

List of Contributors

Barry Allen received his Ph.D in Philosophy from Princeton University. He has taught at the University of Chicago and Hebrew University, and is currently Professor of Philosophy at McMaster University in Hamilton, Ontario. He is author of *Truth in Philosophy* (1993) and *Knowledge and Civilization* (2004), and is an associate editor at the interdisciplinary journal *Common Knowledge*.

James Conant is Professor of Philosophy at the University of Chicago. He has published articles on Kant, Kierkegaard, Nietzsche, Frege, William James, Wittgenstein, Carnap and others.

Diarmuid Costello is Senior Lecturer in the Theory of Art in the School of Arts and Humanities at Oxford Brookes University. He has recently published papers on the aesthetics of Greenberg, Lyotard, and Heidegger, and has forthcoming papers on Danto and Benjamin. He currently holds a Leverhulme Research Fellowship for a monograph entitled 'The Fate of Aesthetics in Contemporary Art Theory'.

Cora Diamond was Kenan Professor of Philosophy, University Professor and Professor of Law at the University of Virginia, where she is now Professor Emerita. She has also taught at Princeton University, Aberdeen University and the University of Sussex. She is the author of *The Realistic Spirit: Wittgenstein, Philosophy, and the Mind* (MIT Press, 1991) and the editor of *Wittgenstein's Lectures on the Foundations of Mathematics, Cambridge, 1939* (Cornell University Press, Ithaca, 1976; University of Chicago Press, 1989).

Daniel D. Hutto is Reader in Philosophy of Psychology and Head of Philosophy at the University of Hertfordshire. He is the co-editor of *Current Issues in Idealism* (1996) and author of *Beyond Physicalism* (2000). He has recently published *Wittgenstein and the End of Philosophy: Neither Theory nor Therapy* (Palgrave, 2003).

Kelly Dean Jolley is Alumni Professor and Chair of the Auburn Philosophy Department. He has published several papers on Frege and Wittgenstein.

Michael Kremer received his Ph.D in philosophy from the University of Pittsburgh in 1986. He taught at the University of Notre Dame from 1986 until 2002. In July of 2002 he joined the department of philosophy of the University of Chicago. He has published articles in philosophical logic, the philosophy of language, and the philosophy of mathematics. Much of his recent work concerns the development of early analytic philosophy, especially through studies of Frege,

Russell and the early Wittgenstein. One of his current goals is to develop themes present in his contribution to *Post-Analytic* Tractatus into a more comprehensive reading of the *Tractatus* as a whole.

Anthony Rudd is Assistant Professor of Philosophy at St Olaf College, Minnesota, and has taught at the Universities of Bristol and Hertfordshire in the UK. He is the author of *Kierkegaard and the Limits of the Ethical* (1993) and co-editor (with John Davenport) of *Kierkegaard after MacIntyre* (2001). He has recently published *Expressing the World: Skepticism, Wittgenstein and Heidegger* (Open Court, 2003).

Walter P. Van Stigt received a Doctorate from the University of London for his thesis on 'Brouwer's Intuitionism'. He is a Research Fellow of Wolfson College, Oxford and has had previous appointments at the University of London, the University of Utrecht, and the University of South Maine. He founded the Brouwer Archive while at the University of Utrecht. His publications include *Brouwer's Intuitionism, From Brouwer to Hilbert* (co-authored with Paolo Mancosu, 1990); 'The Rejected Parts of Brouwer's Dissertation on the Foundations of Mathematics', in *Historia Mathematica* 6, 1979: 395-404; 'L.E.J. Brouwer, Signific Interlude', in *L.E.J. Brouwer Centenary Symposium*, A.S. Troelstra and D. van Dalen (eds), North Holland, Amsterdam, 1981; and 'L.E.J. Brouwer: Intuitionism and Topology' in *Proceedings, Bicentennial Congress of the Wiskundig Genootschap*, part II: 359-374, Mathematisch Centrum Amsterdam, 1982.

Barry Stocker is an Assistant Professor in the Department of Philosophy at Yeditepe University, Istanbul and is an Honorary Research Fellow in the Department of Philosophy at University College London. He is currently writing a monograph, 'Derrida on Deconstruction', for Routledge. He has published, or has forthcoming, journal articles and book chapters on Pascal, Kant, Kierkegaard, Nietzsche, Heidegger, Wittgenstein, Strawson and Derrida in the fields of aesthetics, philosophy and literature, ethics, political theory, metaphysics and language.

Introduction

Barry Stocker

The purpose of *Post-Analytic* Tractatus is to draw attention to the ways of looking at Wittgenstein's *Tractatus Logico-Philosophicus*, which refers to the limits of analytic philosophy, including the incorporation of approaches from Continental European philosophy. The issue of the limits of analytic philosophy, as well as those of ethics and normative political philosophy, has been increasingly addressed in recent years. As a result, analytic philosophy is now seen more as a tradition with a history that necessarily refers to, and may benefit from more in mixing with other traditions.

Whereas *Philosophical Investigations* has previously been seen as the point where Wittgenstein leaves a logicist and positivistic position behind in the discussion of language in itself, the *Tractatus* has been seen as logical positivist, or something comparable, with the addition of some dark mystical comments. Interpretation of Wittgenstein was initially carried out largely through discussion of *Philosophical Investigations* as a text comparable to hermeneutic and deconstructive approaches. This, combined with an interest in continuities between earlier and later Wittgenstein, establishes a way in which to read the *Tractatus*. As a result, an approach to interpreting *Tractatus* has emerged, which is concerned with a precise discussion of saying and showing, sense and nonsense as permeating every proposition. Questions of therapeutic philosophy and philosophical writing, transcendental conditions and limits have come to the fore. The *Tractatus* can more readily be seen as a text concerned with issues in Kant, Schopenhauer, Kierkegaard and others in the history of philosophy, rather than as a clear break with philosophical tradition. The idea of an end to philosophical tradition, or the overcoming of metaphysics, itself has a philosophical history which must be recognized for there to be a proper understanding of philosophical ends. Mysticism itself needs to be understood as something with a proper place in the history of philosophy and the possibility of philosophy, rather than as purely negative category for irrational subjectivity.

This collection of essays gives a cross-section of recent work on the *Tractatus* from those who are already known for their influential *Tractatus* interpretations. Conant and Diamond have defined the field through their discussions of nonsense and therapeutic reading and the presence of issues from Kierkegaard and Frege in the *Tractatus*. Their most recent papers on Wittgenstein are included, along with papers by Costello, Hutto and Jolley, which evaluate Conant and Diamond and continue the discussion of the themes they raise. Papers by Kremer and Rudd discuss solipsism and the limits of expression. Allen and Stocker refer to the Continental European approach from Kierkegaard to Derrida in the *Tractatus*, and the recurrence of metaphysical questions from the beginnings of philosophy.

The collection's first, and longest, paper is not a work of *Tractatus* interpretation, however. Brouwer's 'Life, Art and Mysticism', first published as a small book in 1905, is included on account of its illuminating power with regard to the questions raised in reading the *Tractatus*. It is the work of a major mathematical philosopher dealing with the values of life, the purpose of art and the nature of mysticism. However, as a sustained philosophical argument, it is clearly not work of the highest rank, and most readers are likely to find it bombastic and chauvinistic in many respects. Nevertheless, none of this should detract from the importance of studying 'Life, Art and Mysticism' alongside the *Tractatus*, since it makes explicit what is left outside the limits of the sayable in Wittgenstein. Although Brouwer's view is not identical to that of Wittgenstein, his emphases are those of Wittgenstein: the end of desire or will and the painlessness of contemplation; truth outside science and intellectualism; solipsism; values that lie outside the world; the simplicity of the world; the simplicity of language; the illusoriness of metaphysical constructs; language as that which communicates what is already understood; and the value of naiveté. Through this essay, readers will find a way into mysticism and the ways of relating it to aesthetics and ethics, and the perspectives which may arise from a fascination with logic and maths as models of our experience, even if they don't agree with all of Brouwer's judgements.

Rudd's 'Logic and Ethics as the Limits of the World' follows Brouwer, in order to show the links between the *Tractatus* and solipsistic metaphysics, with reference to Schopenhauer and the issues of how metaphysical truths can be expressed. Like Rudd, Kremer writes on solipsism with regard to the most recent discussions of Wittgenstein in 'To What Extent is Solipsism a Truth?', and his paper is distinguished by a concern with the precise understanding of mysticism. The possibility of reading Frege as a therapeutic philosopher is raised by Jolley in 'Frege at Therapy' suggesting the need for logico-mathematical philosophy to engage with questions of the limits of expression. A continuity between early and late Wittgenstein is emphasized. Costello's '"Making Sense" of Nonsense: Conant and Diamond read Wittgenstein's *Tractatus*' contrasts the therapeutic reading of Wittgenstein in Conant and Diamond with Hacker's 'orthodox' reading, to suggest that the *Tractatus* is an important failure on the way to the successful therapy for philosophical nonsense in *Philosophical Investigations*. The evaluation of Conant and Diamond is continued in Hutto's 'Making Sense of Nonsense: From Logical Form to Forms of Life', where their therapeutic approach is contrasted with a doctrinal approach. Hutto tries to mediate between them, suggesting that a purely therapeutic approach will be self-contradictory. Diamond's 'Saying and Showing: An Example from Anscombe' argues that the *Tractatus* method is one of the clarification of individual propositions, so that it is not so much concerned with what is unsayable as with avoiding nonsense through clear symbolism and use of signs. Conant contrasts resolute and irresolute readings of the *Tractatus* – that is, readings which acknowledge the force of Witttgenstein's suggestion that the *Tractatus* should be read as nonsense, and readings that do not – in 'Why Worry about the *Tractatus*?'. The resolute reading shows that the *Tractatus* changes our relation to our own words, so that we avoid nonsense. Stocker's 'Transcendence and Contradiction in the *Tractatus*' roots the *Tractatus* in Kantian questions of transcendental illusion and its inevitability in reason, and is also concerned with Kierkegaardian strategies of writing, with regard to limits to saying in the *Tractatus*. Finally, in 'Wittgenstein's Onto-Logic', Allen looks at the *Tractatus* as a work of metaphysics in a tradition going back, beyond Kant, to Parmenides and Aristotle, which assimilates the ontology of objects to what can be said or represented in language. The *Philosophical Investigations* is an overcoming or deconstruction of this metaphysical position.

No unified position is offered by these papers which diverge greatly on many issues, including the central recent discussions of resolute versus irresolute, therapeutic versus doctrinal or orthodox readings of the *Tractatus*. Nevertheless all the papers come together in demonstrating that the *Tractatus* is a fundamental text in philosophy's reflection on itself as embedded in propositions and the difficulties of discussing what propositions are, what they refer to and the criteria of propositions, within propositional language. On any reading, the *Tractatus* engages with what it is for language to reflect on itself, and what can be contained within language, thereby transforming our view of Wittgenstein's precursors, all contributions to philosophical tradition and the nature of that tradition. Encompassing Frege and Kierkegaard, Wittgenstein shows how the showing and saying, sense and nonsense are fundamental issues of any possible philosophy. Philosophy cannot be an analytic reduction to basic terms in a single method. Therefore we must be 'post-analytic', recognizing the illusions which arise from trying to found philosophy on a single, unified language of metaphysical, logical or phenomenal concepts, with no room for paradox and inexpressibility.

Istanbul

Chapter 1

Life, Art, and Mysticism[1]

Luitzen Egbertus Jan Brouwer

Translation by Walter P. Van Stigt[2]

I: The Sad World

Holland was created and kept in existence by the sedimentation of the great rivers. There was a natural balance of dunes and deltas, of tides and drainage. Temporary flooding of certain areas of the delta was a part of that balance. And in this land could live and thrive a strong branch of the human race.

But people were not satisfied: in order to regulate or prevent flooding they built dykes along the rivers; they changed the course of rivers to improve drainage or to facilitate travel by water; and they cut down forests. No wonder the subtle balance of Holland became disturbed; the Zuyder Zee was eaten away and the dunes slowly but relentlessly destroyed. No wonder that nowadays even stronger measures and ever more work are needed to save the country from total destruction. What is more surprising: this self-imposed burden is not only accepted as inevitable but has been elevated to a task laid on our shoulders by God or inescapable Fate.

Originally man lived in isolation. Supported by nature, every individual sought to maintain his equilibrium between sinful temptations. That filled the whole of his life; there was no involvement with others, nor was there any worry about the future. As a result hard work did not exist, nor did sorrow, hatred, fear, or lust. But man was not content; he started to assert control over his fellow men and to search for certainty about the future. And so the balance was lost: labour forced on to the oppressed became ever more distressing and the conspiracy of those in power ever more wicked. We have now reached the point where everyone has power but at the same time suffers oppression; the old instinct of separation and isolation now only lives on as pale envy and jealousy.

Animals and human beings originally did not interfere with one another. This happy state ended when these discontented humans started to sponge on the animals which they found useful and tried to exterminate the others. The order of nature was torn apart and turned into misery: the burden and toil of looking after domestic animals; all sorts of disease created by parasitic eating practices; a long and hard battle against the wild animals that had not been exterminated; and an even harder battle against vermin in man's own home and the bacteria infesting his body.

Science takes pride in this battle and even expresses its resignation in God's will, while it is all the result of rebellion against his will!

It is part of the balance of the eternal and omnipresent life that everyone is called away from this earthly existence when one's time has come. Until then man suffers in mind and body as befits his evil mood of thrift, his lust for power, his vanity, and fear. In his resentment he starts tampering with his body through medicines and diets and with his mind through hypnosis and make-believe; he disturbs the melting pot of his lusts and destroys the balance of psychical responsibility and physical well-being. There is a bodily and moral degeneration, such that, in the end, man can no longer be held responsible for his crimes, for what he has done during his time on this earth. Science has recently claimed credit for extending the span of human life, which certainly is much too short. But what is that worth? It is equally sad to leave this life after one's time as before one's time; and as to death, 'Nature never destroys anything without putting something better in its place'.

Meanwhile, truth is still around and about. There is, for example, the nursery rhyme of the little fish in the sea, and such sayings as 'Honesty is the best policy', 'Better is the enemy of good', and 'Truth will be out'. Educators teach little children 'always to tell the truth'; they drum it into them that little lies never pay, that one thing leads to another and that, in the end, one becomes entangled and gets caught. And, of course, there are all those novels, illustrating how, in the end, the chickens will come home to roost. One truth therefore definitely comes through:

> If reason presents certain actions as likely to improve your condition but your conscience does not approve, then leave them undone. Reason never grasps the world in its entirety and the means it dictates to achieving its limited aim will ultimately and in some inscrutable way only cause damage.
>
> If in this life we always had a mirror in front of us in which we could see things at a glance, grasp everything in one image, acting and knowing would not cause us any problem. But since in our viewing we must turn from one thing to another, we cannot concentrate on one without obstructing the other. (Meister Eckhart)

Truth may be around, but the life of each human being and of people as a whole is nothing but a long string of sins against truth. Aspirations are continually frustrated and new ones take their place; all these *castles in the air* collapse and new ones are built in their place.

The life of the individual is an illusion, an anxious and laborious pursuit of ends – disillusionment. At the time of death, which he has awaited unprepared and in complete ignorance, he is either startled by the realization that he has wasted his life, or his reason is dulled by the comforting thought that without illusions life would have been nothing at all, or that on balance at least he will take with him into his grave a large measure of experience.

Oh yes, these 'wise' old people, who kid themselves that experience, old age, a long life of sin which has left its mark on their faces, rigid and long deprived of all naiveté and which stares out of their lifeless eyes, that all this and this alone leads to wisdom! And then, when things come to a head, they challenge the younger generation to tell them what human life is all about.

The life of mankind as a whole is an arrogant tearing up and devouring of its nest on this pure earth, messing up its mothering growth, gnawing and mutilating her and making her rich creative power sterile, until all life has been swallowed up and the human cancer has withered on the barren planet.

The sickness of mind which has caused this, and which has turned men into madmen, they call 'understanding the world'.

II: Turning into Oneself

Having contemplated the sadness of this world, look into yourself. Within you there is a consciousness, a consciousness which continually changes its content. Are you master of these changes? You will probably say 'no', for you find yourself placed in a world which you have not created yourself, and you are bewildered by the unforeseen change and adversity you meet there. But isn't the content of your consciousness in part determined by your own moods and aren't these within your power? Or is the motto 'Control your passions' only an empty phrase? No doubt you sometimes have this religious sensation, when you feel as if you have withdrawn from your passions, from fear and desire, from time and space, and from the whole of this perceptional world. And, finally, you do know that very meaningful phrase 'turning-into-oneself'. You therefore seem to be capable of some kind of attention which centres round yourself and which to some extent is within your power. What this self is, you cannot further say, nor can you reason about it, for you know full well that all speaking and reasoning is an attention at a great distance from the self; that you cannot get closer to the self by means of words or reasoning, but only by this turning-into-yourself as it is given to you.

This turning-into-oneself requires an effort; it seems that some inertia must be overcome, that your attention is strongly inclined to linger where it is, and that the resistance felt in the move toward the self is much greater than in the move away from it.

If, however, you succeed in overcoming all inertia and proceed, you will find that passions will be silenced, you will feel dead to the old world of perception, of time and space, and all other forms of plurality; and your eyes, no longer blindfolded, will be opened to a scene of joyful quiescence.

> When all images have been removed from the soul, and she beholds the Only One, then the naked essence of the soul finds the naked formless Essence of Divine Unity, the presence of the Superior Being waiting in the self. (Meister Eckhart)

> If only for one moment you abandon yourself there where no creatures live you will hear God speak.

> It is within you. If only you can be silent for one hour and forget all your desires and feelings, you will hear the unspeakable words of God.

> When you keep still and let go of the feelings and desires of your self, then eternal hearing, seeing, and speaking will be revealed and God will hear and see in you. Your own hearing, willing, and seeing is a hindrance, stopping you from seeing and hearing God.

> When you are silent you are like God before He formed nature and creatures, including yours; you will then hear and see with what God saw and heard in you before your own willing, seeing, and hearing had begun. (Jakob Boehme)

Then you will understand the content of all your previous awareness, and you will also understand that until then it had to remain incomprehensible to you; you will understand in the sense of being reconciled with it; you will accept it as self-evident. It will make you feel as if you live through it all at once and yet do not live through it in the sense that you do not feel at all bound by it. At the same time, you are also aware of an infinite wealth of fantasies, a mixture of all kinds of worlds, which now claim as much, and also as little, right to existence as those you previously considered to be real. And in this confluent sea of colours, without separation, without firmness and yet without movement, this chaos without disorder, you see a direction, which you follow automatically, yet not of necessity.

You will recognize your *free will*, free insofar that it can withdraw from the world of causality and remain free; it is only then that your will finds a definite direction, which it follows freely and reversibly. Indeed, the self follows its direction steadily and reversibly, and all the fantasies emanating from it have a direction in parallel and they follow it, steadily and reversibly. You will feel free to return when you so wish to the shackles of plurality, separation, time, space, and bodily consciousness. But you do not return, or rather, you do and you do not at the same time. Freely staying outside, at the same time you live your imprisoned bodily life in this human world, live with your shackles, but you are fully aware that you have accepted them in freedom and that they bind you only as long as you wish.

The phenomena succeed each other in time, bound by causality because your coloured view wills this regularity; but through the walls of causality 'miracles' continue to glide and flow, visible only to the free, the enlightened. You will see how, in this imprisoned world, miracles continually break through and how an invisible avenging hand manifestly administers eternal justice. You will also find that over and above physical causality you can see a clear direction in your own life's course, determined by the self and parallel with the direction of the self:

> How this so-called chance is in fact ruled by a firm, wise and wonderful hand; how through your greater wisdom you will live your life in this *sad world* in lasting joyfulness, knowing that:
> 'There is no evil, and no danger, nothing can happen to me.'
> 'I am a child, loved of God, and born to happiness.'

Your journey through this sad world will be a steady passage in a light and colourful cloud, full of love for all that is clear in it, love even for your erring and covetous fellow men, for in your eyes it is no longer a reality separated from the self, but directed from within the self and with the self. You will feel all-powerful, for you desire only that which follows the direction, and mountains will give way to you. You will feel endowed with an all-embracing knowledge; as in all emanations you feel the timeless direction, a unison of past, present, and future within yourself. You will no longer ask what to do, you will do the right thing without any prompting; therefore you will no longer ask for understanding – all will be clear of its own.

Behind everything you shall feel a painless dissatisfaction with yourself, a conviction that all past misery was self-inflicted: you will see how you abandoned the self, and how your shackled consciousness lost its direction; that it had acquired mass and inertia, and that wandering, it followed an irreversible path, driven hither and thither by desire and fear. You will then see how fear and an obsession with saving, born from the illusion of time, and how desire and lust for power, born from the illusion of space, made you attach intrinsic importance to what should only be a fleeting emanation of the self without any reality of its own. And you will see how the false trails of desire and fear led the wanderer to labour, sweat, and toil, to ever new, irreversible changes and to ever greater misery.

With a smile you will look back on the reality of the sad world, a past illusion, and within it your own fear and desire, your labour and pain. But your happiness is no longer disturbed; that, too, is a fantasy without reality, a fantasy of sadness and remembrance.

III: Man's Downfall, Caused by the Intellect

Without pain you now see mankind, cast down by fear and desire, by avarice and lust for power, by time and space, wandering without wings, incapable of lifting itself in self-reflection, chained to the intellect, the spawn of time and space and fossilized in the form of the human head, the symbol of man's fall. Primitive tribes consider headhunting to be a process of cleansing, and took the greatest pleasure in practising it on the most developed people. This is based on the deep philosophical insight that in nature greater differentiation goes hand-in-hand with graver damnation; this insight resides in their hearts, not in their heads.

This highly valued intellect has enabled man, and forced him, to go on living in desire and fear, rather than from a salutary sense of bewilderment take refuge in self-reflection. Intellect has made him forfeit the amazing independence and directness of his rambling images by connecting them with each other rather than with the self. In this way, the intellect made him persist with apparent security in the conviction of a 'reality', which man in his arrogance has made himself and has tied to causality, but in which ultimately he must feel totally powerless.

In this life of lust and desire the intellect renders man the devilish service of linking two images of the imagination as means and end. Once in the grip of desire for one thing he is made to strive after another as a means to that end: for example, in order to change the course of rivers, he builds dams; indulging his jealousy of his neighbour he sets fire to his house; to protect himself against wild animals he builds his house on stilts; to let the sun shine on his house he cuts down trees. Switching attention from ends to means is accompanied by a change in bodily feelings; there is apparently a noticeable change in the bloodstream, which starts in the head. Here too one feels the close connection between head and intellect.

The act aimed at the means, however, always overshoots the mark to some extent; the means has a direction of its own, diverted at an angle, however small, from that of the end. It acts, therefore, not only in the direction of the end, but also in other dimensions. Man's blinkered view prevents him from recognizing the sometimes very detrimental effects of such action but, worse, gradually the end is lost sight of and only the means remains. In this sad world where a clear view of all human activity is no longer possible, a world dominated by drill and imitation – the other offspring of fear and desire – many recognize as an end what was originally only a means. They seek what we might call an *end of second order*; they perhaps again discover a means to this end and again out of line with that end. If this deceptive jump from end to means is repeated several times, it may happen that a direction is pursued which not only deviates into other dimensions but opposes the direction of the original end and therefore counteracts it.

Industry originally supplied its products in order to create in nature the most favourable conditions for human life. But man ignored the fact that in manufacturing these products from nature's resources he interfered with and upset the balance of nature and the human condition, thereby causing damage greater than the advantages these products could ever bring. For example, to meet the demand for timber, man has razed or ruined so many forests that in the temperate climates hardly any edible plants grow in the wild.

And, worse, manufacturing these goods became an end in itself; new industries were called into existence merely to supply the tools to facilitate production, another blow to the balance of nature. Raw materials were recklessly seized from faraway lands, spawning commercial and naval enterprises, which in turn led to moral and physical misery and to the oppression of one people by another.

As the self was left abandoned, the self that knows all about the past and the future, man grew more and more anxious about the future and began to crave for the power to predict its course: thus, science came into being.

Science, which in its original form was wholly subservient to industry, has made up all kinds of general assertions in and about the world of perception. These come true as long as it pleases God; but one day they will suddenly be contradicted by facts and then our scientists will claim 'Oh yes, of course, we always made this or that tacit assumption'. In their incompetence they then set about complicating the issue even further and making so-called corrections and improvements.

But science does not confine itself to serving industry: again the means becomes an end in itself, and science is practised for its own sake. Bodily awareness has strayed so far away that it is all concentrated in the human head, ignoring and excluding the rest of the body. At the same time man becomes convinced of his own existence as an individual and that of a separate and independent world of perception. At this stage, there are radical changes in the direction of man's attention, and these constitute scientific thinking. For scientific thinking is nothing but a fixation of the direction of will within the confines of the head, and a scientific truth is no more than an infatuation of desire living exclusively in the human head.

Every branch of science will therefore run into ever deeper trouble; when it climbs too high it is almost completely shrouded in even greater isolation, where the remembered results of that science take on an independent existence. The 'foundations' of this branch of science are investigated, and that soon becomes a new branch of science. One then begins to search for the foundations of science in general and knock up some 'theory of knowledge'. As they climb higher and higher confusion grows until they are all completely deranged. Some in the end quietly give up; having thought for a long time about the elusive link between the intuiting consciousness (which develops from the perceptional world) and the perceptional world itself (which in turn only exists through and in the forms of the intuiting consciousness) – a confusion which arose from their own sin of constructing a perceptional world – they then plug the hole with the concept of the *ego*, which was self-created with, and at the same time as, their perceptional world, and they say, 'Yes, of course, something must remain incomprehensible, and that something is the ego that comprehends'.

But there are others who do not know when to stop, who keep on and on until they go mad: they grow bald, short-sighted, and fat; their stomachs stop working properly; and, moaning with asthma and indigestion, they fancy that equilibrium is within reach – and almost reached. So much for science, the last flower and ossification of culture.

The standard of living which human civilization has brought lies far below that of the original human condition: worse, what has been achieved has not done anyone any good. Everyone has been left to drag out his life in the environment of one of the service industries. What an environment compared with the original, virginal state which nature offered man in his unspoilt and naked condition! People have dragged each other down in the misery of their culture, which offered victory and power. But we all know that pettiness and cowardly calculation always triumph over heroism and that heroism is nothing but the determination to defy that infernal phrase 'The end justifies the means' and the infernal act of the intellect – the jump from end to means. On the other hand, the intellectual complication generates a counteractive force against the original aim, which is so great that anyone in a state of naiveté who is suddenly faced with a task, be it physical or mental, and who undertakes the task with the full vigour of an uncorrupted body and an uncorrupted mind, will always prove to be better at it than the one with long training and experience. The Boers and the Japanese, who started a war from nothing, did better than the English and the Russians, and Pastor Felke cured more people with common sense and self-confidence than professors of medicine.

These 'cultured' people do not see the wood for the trees; worse, they have forgotten that there is a wood. Anyone who raises the question of the real purpose of life is declared insane by our modern, practical society; but it is only there that such a question makes any sense. But, of course, there is no room for such a profound question in this confined world of desire and fear with its mass suggestion of a system which deems certain things to be desirable for their own sake (such as wine, wealth, love, and even wisdom) and others to be evil in themselves (things like drought, cold, hunger, poverty, as well as murder and adultery). It is a system which society stores up with great difficulty but without success; it flaunts all kinds of needs which, to be satisfied, require hard work and pain, and so frustrate other needs, so that, in the end, all satisfaction remains illusory. Everyone's earthly life ends in great dissatisfaction; death is the collapse and final ruin of the system. Death repudiates the whole of life; it is the violent manifestation of the self in this limited and self-created world, the unavoidable collapse of the Tower of Babel which man in his vanity had built for himself.

This manifestation of the self, however, also occurs before death, during this restricted life, in the various aspects of this system of desire and in the world of perception, which the intellect has created as the carrier of its infatuations, its independent desires and fears. Here it manifests itself in the voice of conscience, in a nostalgic memory of a *paradise lost*, a faint awareness of quiet happiness which was man's original destiny, in a hankering after bliss, religious certainty, and a life of freedom and dedication. All through this sad world this faint hankering becomes a longing and a yearning for the higher, the transcendental.

Conscience, however, when speaking in this restricted world, is silenced. When it penetrates into the enclaved categories, either man's attention is diverted away by strongly felt stimulation and satisfaction of other needs, or it is assimilated by this attention – that is, it is recognized as a need within the closed system and as capable of satisfaction in the system.

Both are a sop to man's conscience and are used by industry; what should have been a cause for heart-searching and penance is turned into an incentive to new endeavours and new pleasures.

Salving man's conscience by diverting his attention is the sole purpose of the pleasure industries and of public entertainment from card games and wine to prose and poetry. Salving man's conscience by recognizing appetites and satisfying them within the closed system is the main purpose of both the arts and religion industries. The self lives in art, poetry, and religion, but is betrayed by its own offspring and put in chains. Music is turned into crude sensual beat and sing-song; poetry must rely on language and rhythm, which are equally base. What should have led him away from desire, pleasure, and fear has become a source of new pleasures and does not cure his addiction to pleasure; just like the beautiful flowers of nature and the beautiful flower of womanhood, they are only admired and desired in order to be plucked and possessed. One marvels at the virgin forests, but only with a view to *cultivating* them.

The Bible condemns the Tower of Babel and, with it, all building and human creation. Yet religion glories in its wonderful temples, a crystallization of man's creativeness and his high aspirations. Instead of banning fear, it offers a faith which appeals to the intellect, a faith which plays on men's fears, a faith which soothes his conscience and also frightens him. Art, which should have been a liberation from the fixed form, has taken on fixed forms everywhere; its main purpose is to make man unlearn everything, but now there are special 'art schools', where art can be learned.

Art and religion in this world are only grand morphine industries; the yearning for a better life is lulled into sleep and reduced to a state of torpidity. All who play a part in the mechanism of society and so help to maintain its evil mass production are kept quiet and happy. In book and drama they are told about reformers, revolutionaries, and recluses, about contempt for law and order, self-denial, freely chosen poverty and hunger, the free life, rejection of the world of perception, indifference to the misfortunes of life, the kingdom of God. Such people and their teachings are greatly admired – at least when presented in writing or on the stage –

but when they appear in real life everyone is outraged and frightened, and they are locked up in prison or in a lunatic asylum. A life full of hardship, danger, and magical forces, under constant threat of death, but a life in which justice and clear conscience conquer all, such a life – which we all deserve and should endure but from which we run away in fear – is banished to the land of fiction and melodrama. There, it is admired, but in everyday reality it is shunned as something gruesome.

Real life demands that we cover up our bodies, our conversation, and social intercourse. It is not decent to show more of oneself than what belongs to the *restricted life*: one's head, one's intellect, and the actions that have a place in society. Neither is it considered decent to want to see more of other people. Intimacy when discovered by a third party arouses feelings of shame.

But self-reflection sees all these dressed-up bodies, lives, and ideas as ugly and abhorrent, as self-contradictory and as caricatures. Everyone, except the Redeemer is a caricature.

Nakedness in the widest sense is only admired in the closed intellect; it is not turned into practice – that is, the long, difficult, and painful road of the soul and of earthly life, full of sorrow and illness; the abandonment first of the intellect and then of all passions, one by one, when every step brings new sorrow and leads to a next step; when only slowly one feels oneself rise again, covered in scars; and when in the end one's nakedness breaks through. On this road there is no standing still; those who start and then stop suffer an imbalance worse than those who did not set off. For example, a vegetarian who continues as a member of the established society is nonsense; such persistence shows an extraordinarily crude instinct, proving that one abstains from meat not out of inner conviction but from some stupid desire or in foolish imitation of others. The fruits of our civilization and power over other human races are closely related to our carnivorous habits; such a vegetarian is therefore a parasite. This half-hearted, parasitic attitude is typical of most vegetarians and of those who practise free love and anarchism. But this kind of social democracy does not arouse the public feelings of loathing that other minor religious teachings suffer.

These immoral and degenerate lives, reflected in the ugliness and sickness of their bodies, these dressed-up, posturing men, these rigid masks of robots, they give the unspoilt instinct a strange feeling of terror. And here, too, the yearning for a better life is being suppressed. The medical industry tries to maintain a quasi-normal balance in the banished human body: the urge to fight and live in the wild is distracted by diets and medicine. In particular, man's craving for the open air is sated by overfeeding; gymnastics and sport soothe the bodily conscience with some sham satisfaction; in spas and sanatoria the *vis medicatrix naturae*, which should have been the archenemy of 'culture', has humbly donned the robe of lackey, in the service of her enslaver and master. The medical industry was in good hands with barbers and quacks; practised within the confines of the intellect, as a medical science, it is far less effective.

Within the closed system of science the manifestation of the self creates needs, and within that system these needs will be satisfied. In science, too, there is a yearning for something higher, but a yearning which is appeased with religious doctrines of revelation, with metaphysics, ethics, philosophy of art, spiritualism, and theosophy; they all leave man in the sinful bonds of science, of belief in 'reality' and of logical thinking. Here too, instead of escape from earthly shackles, there is a growing insensitivity and a sham equilibrium, bought with ever greater complication of human needs, ever deteriorating living conditions, ever more backbreaking work, going ever further adrift.

Every now and then conscience breaks away from the bonds of the sad world. For example, around the age of eighteen many people discover in themselves a pure, central, and not merely aesthetic admiration for dreamers, monks, and hermits. Some can do nothing but refuse to bow to what middle-of-the-road sages describe as 'life'; they must express their heartfelt (i.e., felt in the heart, not in the head) contempt for all the fruits of culture, for all collaborators in the social chaos, for all those involved in the building of the Tower of Babel, for all those talented tightrope dancers and magicians who are proud of what should be a source of shame, for the social do-gooders and reformers of all hues, who seem to think that God has created us in order to improve his work.

But sadly this free conscience is short-lived: education appears on the horizon and throttles it in its web. At first they learn that there is no work to be done, that there is nothing beautiful or important; they then begin to search for what is more beautiful or more important than the human condition demands, and finally they bow down even further – they become lackeys in the palace of evil, lackeys with their crawling servility toward their master and cruelty toward outsiders, doing ignoble, degrading work, brazenly scrounging and at the same time fearing for their own dear lives.

IV: Atonement

This corrupt world, as you now recognize, only exists because of its very corruption, its deviation from the paths of rectitude. A world of righteousness now seems to you as contradictory as your own mortality. Folly and misfortune, equally balanced, they govern the world! Any struggle for a better order is just one drop more in the ocean of folly. In a world which you know to be 'real', strife and conflict of irreconcilable interests are essential, and so is the search for an external balance which is incompatible with external existence. Any attempt to eliminate this imbalance can only shift the imbalance to somewhere else. An external, visible world necessarily lives on the illusion of a free will; that is where it tries to find happiness, even if this free will remains firmly locked up in causality. Therefore any emanation of power, any strong manifestation of life, and any flowering and growth which are willed shall come about, but only to fade away in spite of strenuous efforts, and the growth at some time in the past will be stemmed. All that is achieved on this earth can be served up in the two acts of the short tragedy called *Grandeur et Decadence*, and in the words of the mystic:

> In God's wisdom it has been ordained
> that man must part from what is dearest to him.

Knowing this, you become reconciled with the erring world and accept its disconsolateness as natural; moreover, you feel it to be your inescapable karma, to which you have reconciled yourself and which you must fulfil, to see yourself driven away from the self, placed in life where pain and labour, desire and fear are your share and where all truth is veiled.

You look on this life as the direction of your duty, and you live it as directed from within the self: in other words, you recognize that all these earthly bonds remain your inevitable karma until God releases you. No new desires will be able to deflect you from your path and you will not wantonly increase the burden of your karma. On the other hand, you will not try to be better than you are because that too would be surrender to evil desire; neither will you wish the world to be better than it is because that would be evil lust for power. Instead you will say, 'What is a God who does not become flesh in a sad world?'

> The zest for natural life of divine power from which spring nature and free will makes one yearn for release from one's own natural [i.e., worldly] will. This same zest is implanted in your will together with the imprint of nature, so that with it God be given a place therein. At the end of time it will be freed from the vanity of nature and be reborn in a crystal clear, pure nature. It will then become clear why God had locked it in time and subjected it to pain and suffering, so that through natural pain one would come to know his eternal power through forms, shape, and mortality, and so that in this time life would be revealed, although in a created form, a countermove in the game of his divine wisdom. Wisdom will be revealed through folly. Although folly will lay claim to it, it has its origin and beginning elsewhere. In this way eternal life is demonstrated through folly, so that folly contributes to the glory of God and the eternal and permanent are known through what is passing and mortal.
>
> In order that the eternally happy may know their real selves, the pain of their earthly suffering, the possibility – though not the reality of a different existence and of downfall – must be the source of their joy, and darkness be a manifestation of light so that light be revealed in experience, which would not be possible in the *One* [i.e., the closed self which one had never left]. By contrariety one will learn what love and suffering are. (Jakob Boehme)

Then you will be reconciled with your world and not try to change it. You will work, eat, sleep, and travel in your world, knowing it to be your inevitable karma. It is precisely this awareness, your humility, which will help you grow in the fullness of the Lord, who will protect you from desires and fears which are not part of the task given to you.

V: Language

The immediate companion of the intellect is language. From life in the intellect follows the impossibility of any form of direct communication with others – instinctively by gesture or looks, or even more spiritually through the separation of distance. People therefore start training themselves and their offspring in some crude sign language, painfully and with little success, for never has anyone been able to communicate with others, soul to soul. Language can only be the accompaniment of an already existing mutual understanding. Even when two people share the same needs and aspirations, they will be in constant danger of being led by their uncontrolled desires into different side roads and of drifting apart; they will suffer pain and anxiety in their struggle to keep together. Only in the very narrowly restricted domains of the imagination such as in the exclusively intellectual sciences – which are completely separated from the world of perception and therefore touch the least upon the essentially human – only there can mutual understanding be maintained for some time. There is little scope for misunderstanding notions such as 'equal' and 'triangle', but even then two different people will never feel them in exactly the same way. Even in the case of the most restricted sciences, logic and mathematics – a sharp distinction between these two is hardly possible – no two different people will have the same conception of the fundamental concepts on which these two sciences are constructed; and yet their wills are parallel, and in both there is a small, unimportant part of the brain which forces their attention in the same way. This also happens when people together fight a common enemy, together build a house or bridge, go into business or strike a deal. Then, too, language will serve its purpose: that is, to keep the wills of separate people on one path.

But ridiculous is the use of language when one tries to express subtle nuances of will which are not part of the living reality of those concerned – when, for example, so-called philosophers or metaphysicians discuss among themselves morality, God, consciousness, immortality, or the free will. These people do not even love each other, let alone share the same subtle movements of the soul; sometimes they even do not know each other personally. They either talk at cross-purposes or each builds his own little logical system which lacks any connection with reality. For logic is life in the human brain; it may accompany life outside the brain but it can never guide it by virtue of its own power. Indeed, if there is a harmony of will, logic may well fall by the wayside; for example, the simultaneous pronouncements 'There is no evil' and 'There is nothing but evil' may well express 'unity of meaning'.

Ridiculous, too, is the use of language when there is an argument and people try to come to an agreement by means of reasoning. Both parties are so much under the influence of society's mass suggestion that they feel ashamed of appearing to be 'unreasonable' – that is, to admit that they search for something different from 'the good' and 'the right', that mirage of human society. In this case language, which presumes a harmony of will, may well be used to accompany strife and combat. But they might just as well keep silent; they only play off their wills against each other and work on each other's desires and fears, and the strongest man wins.

Ridiculous also is the language of conversation. Everyone waffles, but it is considered an art to waffle without nonsense showing through the restraints of convention which hold society together; to score off others, expose their stupidity while keeping oneself covered and within the bounds, and yet daring to touch upon the most subtle topics, that is considered really great talent, demanding the greatest respect – especially in France – and entitling one to be known as *spirituel*! Admittedly, such mock battles are to be preferred to the quasi-serious debates on art and politics.

Comical is the language of conversation between boys and girls. In their case there is already a harmony of will, and language is completely superfluous; indeed, its only purpose is to hide this harmony of will, serving modesty and shame which dare not face it openly, to mask seriousness behind jokes. Seriousness in such conversation is only then acceptable if forced togetherness makes a dutiful exchange of a few words otherwise unavoidable; if one allows any seriousness to creep in between the sexes, all noble modesty is lost. Once you have given away seriousness, you no longer have all to give, although often mock seriousness, a kind of playful coquetry, is the only means of defending one's purity against uncivilized intruders. The worst and most disgusting case of such uncivilized intrusion is that of associations – one of them of students at Amsterdam University – where members, male and female together, 'study the problem of sexuality'. The association in question calls itself *Ethos*, and it represents the greatest obscenity that one has dared to show in public! That such is possible in our modern society just shows how deep the human sense of criticism has been buried in the intellect, far removed from all central instincts.

In everyday life, language only makes sense as a means of holding the already harmonious wills of two people together on one path. The belief in a reality, the same for all – existing outside and independent of them – made society foolishly attach great importance to 'speaking the truth'. Yet 'telling the truth' is often far less effective than what is known as 'telling a lie'. Once someone is imprisoned in the belief in a logically coherent (i.e., conceived without pain in a certain region of the brain) complex of externalities, which he calls 'reality', it becomes rather difficult to follow him in his folly, and even more difficult to try to invoke in him a particular emotion or state of mind by means of words which he can only interpret in accordance with his reality. One would at least have to resort to some gross exaggeration. For example, the subtle teasing and the fun a man pokes at his wife would not mean much to an outsider; they may become a little clearer, if he is told that certain events took place, which did not actually happen, but which might have been a natural consequence of the relationship in question. It is hard for the attention to break away from the intellect – so much so that only the most extraordinary events make an impression and get through to the central human emotions. By using language as the slave of an illusion of reality one cannot reveal truth.

In a similar way does not the *comédie de caractère* – and also naturalism – try to pass off a vision of the world as reality by exaggeration or by pure invention? And do not paintings differ from photographs of nature in the same way?

Language can accompany man's will to dominate the will of others or his will to keep the movements of wills together; for example, the war cry of Red Indians accompanies the will to break the will of others.

Language by itself has no meaning: any philosophy which searched for a firm foundation based on that presumption has come to grief; lulled into sleep by the assurance of such firm foundation, one was rudely awakened by the later appearance of deficiencies and contradictions. A language which does not derive its certainty from the human will but which claims to live on in the 'pure concept' is an absurdity. It is indeed a great skill to be able to go on speaking without being caught in contradiction or without making silent presumptions rooted in the will – sophistical reasoning which requires the brainpower of a man like Bolland – but a kind of skill one admires in an acrobat. Mr Bolland has shown that it is possible to speak within the confinement of reason, to remove language from the sovereignty of passion and emotion – where it originated like all other expressions of life – and to do that without going mad or becoming sick. Physiologists have shown that a frog's organs can be kept alive even when cut off from other organs, but the heart of that frog keeps going for a relatively short time, like the philosophy of Mr Bolland, who maintains it is only his Sunday suit. But when reasoning touches on live issues such as love, nature, and politics, then it makes pronouncements which are lifeless – that is, have no meaning for life.

Language only lives in and through human culture, which on the one hand needs mutual understanding but, on the other hand, makes direct communication impossible. The use of language also consolidates that culture since it operates in the same sphere. People who use language lose their primitive desires which, however sinful, remain close to the self. Frightened by solitude, their only home, they become automata, slaves of the monster-machine of public relations. Their attention becomes shut off from all other influences and from any communication soul to soul. If these influences manage to break through the constraints of their world of intellectual perception with its man-made natural laws, they try first to ignore them and, if that does not work, to study and categorize them, to bring them within the categories of their highly acclaimed 'science'. They do not seem to realize that the purest reaction to such influences is simply to keep a completely open mind, without any prejudice. Even the simplest, everyday jobs are often best done as part of a thoughtless grind of daily routine rather than from some studied conviction. Influences which do not necessarily relate to this everyday life should, in any case, be considered as concealed from our understanding in accordance with God's will. Only in this way can we trust our actions and our knowledge. Tiresias and Cassandra were not members of an association for psychological research: they saw the future when this was necessary; they did not desire this insight nor did they make any particular effort to achieve it.

Modern science, however, does not consider anything sacred. Once a particular influence is observed it must be researched and be relegated to the index of old intellectual categories; answers must be given to the questions 'how old?', 'how far?', 'how big?', 'how strong?', and 'how much does it cost?'. But he who can still free his feelings from the straitjacket of public convention, and who has developed a more delicate sense of perception, nurtures and reveres it rather than dismisses it as 'beside the point'; he will place his trust in dreams and premonitions without wanting to understand them. He will understand the signs given to him without using the power of his 'head'; he can tell the character of his fellow men from their faces, or perhaps more easily and directly from their hands, which do not wear the mask of comedy and coquetry. He will see the most famous and learned men, their self-satisfaction written all over their faces, hailed, admired, and carried high by fools; he will see them naked and stripped of all their glory by just glancing at their hands. Even the most clever orators and philosophers, whose words may seem irrefutable, are given away by their hands.

Such insight, however, is not given to those who have made a study of Lavater's *Physiognomy* or whose intuitive perception has been impaired by intellectual considerations. Only he who humbly opens his free senses, only he will always be guided through life by timely premonitions and apparitions, not those who do scientific research into telepathy or spiritualism or take part in that sort of séance or show. Theosophists and their ilk, who are so keen to find out more about life after death, will receive a nasty shock once they get there.

People who try to force these matters into some kind of science will probably succeed because they will this interpretation. They have cast aside all meekness and innocence and think they have found refuge in some balance: it is a sham balance which again and again will be destroyed by the discovery of new phenomena. Bodily work on this earth will become ever harder and more complex, and so will the search and reasoning of the intellect. Faith, however, will spite gravity and mass and always literally 'move mountains' and walk across the sea. Retiring into the self we will playfully break all laws of nature'.

VI: Immanent Truth

The manifestations of the self within the restrictions and in the forms of this life are eruptions of truth. Always and everywhere truth is in the air and, wherever it breaks through, truth is always the same to those who understand.

When it does break through, truth points to a life where the self, never abandoned, has been found again; where man accepts his earthly shackles in all humility, fully conscious of the inevitable karma of this sad world and his own individual place in it.

And yet truth itself cannot help find the self again; that can only be done by what transcends the forms of this world and what mystics refer to as 'divine grace'.

Truth which in this world points to the inevitability of the karma of that world, which through all the restless move of human desire reveals eternal justice, which points to the obvious collision of conflicting and irreconcilable interests and guides man away from appearances – that is, creations of his own imprisoned desire – such truth is *immanent truth*.

Truth which guides man in this world toward a personal life, free from the shackles of fear and desire, a life where the wisdom and bliss and the quiet jubilation of self-reflection are the fruit of humility, poverty, quiet devotion to earthly duties – his own inevitable karma – this truth is *transcendent truth*.

Immanent truth enlightens; transcendent truth makes man devout. Immanent truth sees the 'idea' of the world. From the viewpoint of so-called reality it may appear to be a lie or at least an exaggeration because of the false forms in which it must dress itself, especially in literature and the 'visual' arts. It conflicts with current opinions which have all grown out of a world-view based on outward appearance – that is, appeal to human desire. And yet it is tolerated only if it can be made to fit into the confined life without upsetting its structure. It is found in music, which appeals to the senses still unaffected by the intellect, to a lesser extent in the visual arts as was shown by Heinze, and least of all in literature which addresses itself directly to the intellect, life itself. The latter is duty-bound to act as the obedient servant of the lie that is human culture, to be enjoyed as a kind of uplifting, ennobling, or edifying experience, but not to be taken seriously in its claim to change the view of the world. Dead authors do not seem to address themselves so directly to the living will of the reader as do contemporary writers. The latter become literary figures only if they become workers in the industry of conscience-salving mock edification or titillation; their working material may even be truth, but only truth dressed up in the fashionable clothing of the prevailing cultural system. Later, however, when the cultural system changes, their clothing is no longer fashionable and they do not even survive as dead authors. A contemporary author is never forgiven for telling the naked truth; the work of the dead is covered with the conciliatory veil of unreality, and their naked truths are swallowed as some vague sacred doctrine.

One puts up with truth in verse more readily than with truth in prose, because poetry is the garb of the courtesan, appealing to the lowest sensibility of the time-bound intellect, its sense of rhythm; when she gives her all, carried on the beat of 'rub-a-dub, rub-a-dub' ('tommy-rot, tommy-rot'), she gives the impression of not to believing a word she says. When you hear a poem such as

La vie est vaine
Un peu d'amour, un peu de haine,
Et puis bonjour

you get the feeling of a whim or a mood which seems to thrive in our culture like so many others, but not a deeply felt truth which challenges that culture.

And, again, speaking the truth in serious conversation is more dangerous than in books or on the stage, as is well appreciated by all who value their lives. As to the truth preached from the pulpit, people listen quietly because it sounds so unreal: the parson preaches that it is sinful to worry about tomorrow and yet he takes out an insurance policy on his house against fire and burglary. Perhaps nowhere else can one hear more truth spoken than in our churches, but nowhere else is it more reduced to something to be heard but not to be practised.

Art, which is real truth, belies common sense, causality, and science everywhere; it kills the optimism which props up the folly of this earthly show; it sees the avenging of fate in everyone's life, how the illusion, the hope, and trust in the stability of this world is turned into misery, spiting the illusion of causality; it recognizes the plurality of this world, attaching to every part a separate and encaged will which never finds rest and is always frustrated by the opposing will of other parts from which it remains separated. At a time when one only believes in knowledge of the intellect and in the natural laws of practical everyday life, immanent truth will continue unperturbed to speak in art of magic, premonition, murder-by-thought, resurrection, healing-by-love, apparitions, and heavenly messengers; it shows men dying, not because of blood poisoning, tuberculosis, or gout, but simply because their time has come, and it does not consider dying crushed under a falling tree less worthy than dying of a stroke.

The naturalist conception of art is therefore not concerned with real truth. Art, according to Zola, is a description of nature as seen by *individual temperament*; but temperament is no more than a titillation of the imagination into some crude frenzy, not to be rated higher than the sentiments of a Sunday night audience of a melodrama. What then remains of nature is nothing but a piece of the outer shell of this world, mainly of its human society, seen as an aseic physical phenomenon under the influence of causality. What remains as the only alternative is some more or less regulated *historic materialism*, a folly of science, but not truth.

Molière's satirical portrayal of human desires, weaknesses, stupidity, and impotence is only the negative side of truth; it disturbs the customary view of society which looks at fellow creatures with optimism, appreciation, and perhaps with fear. But what is positively put in its place remains a senseless, petty and incomprehensible game of appearances, a 'comedy' in the worst possible sense, no better than the astronomers' views of the great cosmic events.

The naked immanent truth bears no relation to the present situation, the prevailing cultural system. Art, which is truth, is of all times.

Distinctions can be made as to the extent to which the self-destruction of the illusion of time or of space is revealed. The former is clearly shown in music, but also and more fully – though less forcefully – in literature, particularly when it considers time from the viewpoint of a frozen *now*, as it does in drama. Epics are narratives, they leave in the reader an awareness of the separation of time and therefore stop and dwell on *externalities*. Comedy, even when it freezes the present, does not step out of time but continues to live in parallel with time; it stays within a fading flow of time. Its denials are not convincing; they centre on externalities and arouse nothing but feelings of excitement. Tragedy, on the other hand, enters a static present and withdraws from life; it recognizes life as a continuing creation from illusion and a decomposition, an illusion and a cruel disenchantment of fate, whose clouding wings spread over this earth and quash any rise above man's immutable karma and sling him back into the mud. Fate's cruel humiliation is accepted as justified and inevitable, as is man's thwarted attempt to raise himself. A world contented and resigned would serve no purpose; karma wants to rise above itself and will always be forced back into itself.

In the tragedies of Sophocles and Shakespeare, the protagonists, Oedipus, King Lear, and Julius Caesar, follow the paths of their fate wandering in darkness, but one can sense the outcomes right from the start. In *Hamlet* illusion and disillusionment are so closely allied that they always appear together. In the course of the play the hero dies many deaths; whenever he reaches for a firm hold it is snatched from him and again and again he is forced back into the flow of his karma. Death finally comes; its tragic justice as the disavowal of life is an indispensable part of all good tragedy. At the end of the play Hamlet must die, every single illusion has been shattered – happiness, trust, and love – then why not the sum total of all illusions, life itself? That is also why, in *King Lear*, Cordelia, who had never done any wrong, must pay with her life just as her evil sisters had. Everything positive in this life, every act, every personal quality good or bad, will punish itself in cruel death – cruel because it is felt as an affliction, whether or not it is feared in advance and suffered in pain during one's last hour. In this sad world everyone commits certain acts and has certain qualities and therefore lives in the original sin of his birth and in the expectation of painful atonement.

The visual arts lack the flowing element of time and therefore cannot reveal the self-destruction of the illusion of time but they do reveal more deeply and more directly than drama the self-destruction of the illusion of space, the illusion of plurality which is already receiving its punishment, the pain of gazing, baffled and helpless at this plurality, a pain which one tries to escape in a never-satisfied desire for possession – that is, joining it to the individual self, hopelessly forsaken and helpless. Straying further and further away, attention centres on the world outside and so adds to the burden of karma: it becomes lust for power, for money, for glory and for … the illusion which is woman. The latter indeed is a burdening of karma, for in the full karma of man there is no room for woman: she is a Siren luring him away from his path.

There is a balance between man's burden of guilt and the burden of labour and toil imposed on him. A similar balance is found between woman's wantonness, her inborn capacity for karma burdening, and the measure of femininity which this world offers in temptation. In a world of humble acceptance of given karma there would be no women. But again such a world would serve no purpose, the wantonness of this world is inseparably bound up with its continued existence and its sufferance of womanhood and, amazingly, the latter are also empirically found to be inseparable. It is a particularly strong case of the different, ever conflicting, and irreconcilable interests in this world of plurality: man must shun and ignore woman in order to avoid increasing the burden of his karma and his own ultimate downfall – listen to Shakespeare's Anthony under the spell of Cleopatra, crying out in desperation 'I must from this enchanting queen break off' – and woman cannot exist without man, her whole karma is nothing but her sex, so much so that between a woman and a lioness there is less difference than between twin brothers.

Woman must live in a world where she feels everything but cannot be anything. In her body she experiences the feelings of humanity, of race and family, but she may not indulge in them. Only one thing is left to her: one who is her ideal, one whom she may follow with her eyes without asking of him anything for herself, no love in return, not even being noticed by him. She is to be an instrument, sent from heaven to loosen the bonds of his karma and help him keep away from disturbing temptation. But in doing so she does not notice that her entrance into his life becomes his greatest temptation, when he begins to feel an all-giving love toward her. In her consciousness, in her attention she helps him live his life purely in himself, but underneath, from the dark depths of her sex, she will lure him on to paths which lead him to ruin.

She shall be humble, and humbly she will want to take from his hands all ignoble work, all work other than the pure enjoyment of the faculties of the body in which he walks this earth. Without flinching, she will give her life to save his equilibrium.

Her look shall be serene and calming; she shall live her life doing anything for her beloved with dogged persistence and patience, her body unwrinkled and unmoved, without desire to seduce and unaware of its seductive power, and yet a body so irresistibly tempting in its tormenting repose that no man can resist.

The Venus of Milo symbolizes the karma of woman, the still, passionless woman, so unaware and yet so devilishly seductive. However, pure female love, love without temptation, such serene untroubled love as can be found between brother and sister, can be wonderful.

In the meantime, woman will sin and burden her karma, as indeed man does; she will do this through her female passion for her beloved and also by engaging in male activity. An example of the first is Gretchen's monologue in *Faust*.

My peace is gone,
My heart is sore:
I never shall find it,
A nevermore.
Save I have him near,
The grave is here,
The world is gall
And bitterness all.

My poor weak head
is racked and crazed.
My thought is lost,
My senses mazed.

My peace is gone,
My heart is sore:
I never shall find it,
A nevermore.

To see him, him only,
At the pane I sit
To meet him, him only
The house I quit.

His lofty gait,
His noble size,
The smile of his mouth,
The power of his eyes.

And the magic flow
Of his talk, the bliss
In the clasp of his hand,
And ah! his kiss

My peace is gone,
My heart is sore:
I never shall find it,
A nevermore.

My bosom yearns
For him alone;
Ah! dared I clasp him,
And hold, and own!

And kiss his mouth
To heart's desire,
And on his kisses
At last expire.
(*Faust*, trans. B. Taylor, OUP)

Female passion is totally different from male passion: it is free from the illusion of space and therefore does not seek satisfaction in the owning of property. It is a blind fantasy within her, often punished by turning eventually into a loathing of the man she once desired and yet without the power to stop her desires for him.

The sin of engaging in male activity – living the ideas expressed in her body and ignoring her femininity, which does not approve – is now sanctioned by the perverse doctrines of modern critics. One can nowadays even mutter with impunity a slogan such as 'Man and woman are equal'. And whatever human folly wants will happen; perhaps all work that is considered man's prerogative will, in the future, also be done by women, maybe even exclusively by women. Yet human folly will be unable to alter the general karma of this world; this will always be the same: work that in the prevailing creed of race and culture is deemed to be noble will remain the prerogative of man, and the ignoble, mundane tasks will as much as possible be done by women. The gradual usurping by women of certain work will inevitably lead to the degradation of that work. People's views as to what is 'noble' work changes with the times; in the age of chivalry it was fighting and hunting, later it became politics, in modern times it is science, especially the kind of science as is practised in universities. Such noble work has always been reserved for men; women were excluded. Two recent examples of change or so-called development have been: the debasement of universities into places where wage-earners are trained in disagreeable, wretchedly necessary but degrading social work; and the admission of women to these establishments. Until recently the state, public life, was considered to be an honourable institution, even something metaphysical. Work in the public sector ranked high; it was a noble task in contrast with domestic work, which is necessary but wretched and inferior. During the last century socialist movements swept away that noble and honourable status; at the same time as women started to take up positions in public life, first only in a subordinate administrative role. The management of great enterprises still relies on male passion and male folly; but when, at the end of this socialist process of decay, the state is no more than a well-oiled robot, then perhaps the whole of its administration will be left to women. That money for one's livelihood is usually earned by the man is of as little importance as money itself. This happens to be so at present, when earning money goes hand-in-hand with doing noble work. The old Germanic tribes regarded tilling the land as ignoble, inferior work and it was, therefore, done exclusively by women. When all productive labour has been made dull by socialism it will be done exclusively by women.

In the meantime, men will occupy their time according to their abilities and aptitudes in sport, gymnastics, fighting, studying philosophy, gardening, woodcarving, travelling, training animals, and anything which at the time is considered to be noble work, even gambling away what their wives have earned; for even that is much nobler than building bridges or digging mines.

In this way the sin of male activity becomes a hopeless struggle of woman against fate, which had allotted her nothing but menial, ignoble tasks. This carries with it its own punishment in the uncomfortable feeling of never experiencing within herself the strength-giving drive to do this male work and never understanding the work she is doing, however good she may be at it. No matter what male activity she engages in – whether the simple realization and expression of the male idea or its more frivolous aberrations – her sin remains the same: amazons, female writers, and painters are no better than female butchers. A philanthropic woman is just as much a caricature as a cruel or ambitious woman.

If a woman can keep herself free from passion and activity, she will still experience and feel the shackles that imprison her nature as an atonement for past guilt, failing to know or find her ideal. Groping in the dark and childlike, she will at first admire and share the minor male talents and fantasies, unable to do more than share in and agree with the peripheral sentiments of a man. Only a few will go on to see the whole of individual man, including his fate, and only this can be called 'love'. She may then understand his fate and his life even better than he does himself, and in his aberrations she must suffer the pain of not being able to hold him in as high regard as she would wish. If his fall away from his karma is permanent and not toward her (a fall toward her would be the only one she would not recognize), then his fall away from himself would also be a fall away from her. She then must surrender all that had any meaning in her life, and in doing so she acts in accordance with her duty. Clinging to him in desperation would be typically male tenacity. Real love does not survive contempt. She shall bear her loneliness with patience until one day a new and higher male sphere is opened to her, a less burdensome male karma. There will be lovers, one after another, and every time she will let her beloved go when he falls away from his karma permanently, or when another love, one of a higher sphere, is revealed to her. Only in her life and by living will her ideal become clear to her: the highest male principle, which surpasses fear and desire, which cannot fall away from karma because it is above karma, which is not concerned with power or talent nor with good looks or character, but which is nothing but humble courage and clear vision. To know this ideal and yet not be able to find it in this world, that is her ultimate torment during the whole of her life.

But this is not the life pattern of most women, because woman, too, strays away from her karma. First it is her female passion which makes her want to draw her beloved to her; it is felt as an emptiness to be filled by him. Then it is male activity which burdens her body, creates an ideal which is not the highest male principle but some talent or other, usually a talent which accords with her own nature. The latter is in contrast with man, who wants to possess what in his ignorance he feels to be different, outside himself, and who is most easily tempted by and drawn to the opposite of his own nature, another example of the conflict of interests in this world. Gretchen in *Faust* expresses her admiration of such talent of male activity.

Dear God! Such a man,
Who can do all and everything!
I stand before him in shame
and say yes to all he says and does.

I am only a poor and ignorant child
who does not understand what he sees in me.

But woman strays further and further away; overcome by ambition, fear, and jealousy she is drawn away from her ideal to other men. If she has taken on the male characteristics of self-consciousness and ambition she will become a temptress of men and so trade in her womanly ideal for the lowest in the make-up of man.

Such is the inevitable, sad state of love in this world; the pure forms of love will only come to life as the world and she herself are destroyed.

But truth in art shows in never fading lines that man should shun and ignore woman but that woman should live in man, consider herself to be nothing, powerless and worthless, and should sacrifice all for her beloved. A true woman is pale, smooth, and without expression; her eyes are dull and dreamy; she has no physical strength and yet she shrinks from nothing. But any man who turns to a woman loses his life. It is the old story, briefly described in a vision of Marie Madeleine.

I had a dream and saw a tree,
Youthful and full of the strength of spring;
And in this dream I saw a tropical flower,
Winding itself around his bark and drinking his sap.

She was very white, and she never weakened
Taking the sunlight from the other's face.
She drank his blood and sapped his strength.
The tree withered. – It was only a dream.

No work of art portraying love is hailed as great and true unless the woman, and she alone, is a most splendid figure; the man is usually described as a poor duffer, completely thrown off-balance. In *Hamlet*, the truest of all drama, the hero also represents this aspect of male karma; despite all the love he feels for Ophelia, the seductive temptation in which he feels trapped, his conscience does not allow him to let himself go! But, in her case, all attention is centred on his fate, his sadness and confusion, the fateful course on which his life is set.

Male love is always portrayed as frivolous and as a sad, blind passion, while womanly love is raised to the heights of sublime fate. The idea of human love is the subject of Shakespeare's *Anthony and Cleopatra*. She shares life in the highest form of which she is capable and as she finds it expressed in her beloved; but he – for this very reason – is lured away from his right path and squanders all that is noble in him for her sake. His life is ruined, and after his death she too parts from her life that now has lost all meaning. The burning of widows was once a sacred rite, but it was banned by Western barbarian governments as barbaric. Adelbert von Chamisso gave a pure rendering of womanly love. In his *Song of the Three Sisters*, two of the sisters tell of the suffering their loves have brought, but the greatest pain is that of the third sister who sighs, 'only a few words: "I have never been loved..."'. She should have said, 'I have never loved'. Only through love does she become woman, but in doing so she loses her own identity, as she admits when she says,

> Startled and ravished by my Friend,
> I am lost to myself.

See how little desire she has to draw him toward her and to bind her life to his:

> Wander and go your way,
> I shall only see your shadow,
> Watch your image in all humility,
> And be happy and yet, sad.

Also:

> Don't listen to my quiet prayer,
> Only devoted to your happiness...

This is because sublime love goes hand-in-hand with a deep sense of shame, an instinctive shunning in his presence of the temptation which emanates from her; and no matter whether he falls under her spell or not, even knowing her is a distraction to him. Important to her happiness is not what she does or what happens to her, only his life matters and what happens to him. The health of the ideal woman is not affected by her own diet but only by that of her beloved. Physically, too, she literally only lives on love. She is cured of every illness by the mere touch of his hand, by his breath, but she does not have a similar power over him.

Since she lives in nothing but love, she does not have any desire for, nor does she feel herself capable of leading a life of her own. She does not know human – that is, male – desires; moderation and abstemiousness are typically female qualities. As to worldly aspirations and political convictions, she will simply and naively follow in the footsteps of her beloved, adopt his opinions without question and defend them against others as if they were objective and unassailable axioms. Disputes with women clearly show the ridiculousness of language as a means of reaching agreement and the notorious phenomenon of *female logic*. Goethe speaks of

> ...these women, who after hours of reasoning,
> keep going back to their first sentence.

Immanent truth breaks through even in science. Science places whatever is perceived, outside the self, in a world of perception independent of the self; the bond with the self, its only source and guide, is lost. It then constructs a mathematical–logical substratum which is completely alien to life, an illusion, one which acts in life as a Tower of Babel with its confusion of tongues. But in self-reflection man sees the world surrounding him as his karma bearing his own guilt and the confusion in this world, caused by his activity and reasoning, as a reckless and self-inflicted aggravation of that karma. He will withdraw from it all and no longer collude in this arrogant interference with nature, the wilful evocation of phenomena which seems to be the main preoccupation of the physical sciences. Whatever is perceived as independent of his own action will be felt and seen in a kind of polarization as an image of his own fate; the true self will accept it and live with it as something free and yet something obviously necessary. Living in what he so beholds as the one pole of this polarization, he will not lose the bond with the other, the source of permanent tranquillity and wisdom. The blue, firm sky will be felt as the exact antipole of his own mood of humility and contemplation, the firm course of the stars as the antipole of his own freedom, the colours and branches of plants as the antipole of yet other colours and of the passions in his own blood.

These insights break through as immanent truth in the science of culture. Once alchemy and astrology were cases of such a disturbing breakthrough; modern chemistry and astronomy are just slaves of culture like any other branch of natural science. In any of these cases, however, the breakthrough of truth always moves the centre of gravity back again from the observed to the observer: Copernicus moved the rotation of heavenly bodies down to earth, and one day it may well be placed in man's own body. Kant replaced the study of the properties of things by that of the human head, man becoming aware of categories. Positive, quantitative properties are again and again replaced by polar ones; for example, in the new theories of electricity and light, Newton's theory of colour analyzed light rays in their medium, but Goethe and Schopenhauer, more sensitive to truth, considered colour to be the polar splitting by the human eye.

Of course, none of this really matters; it leaves the world as stupid as before. It is not what we described as turning-into-the-self, turning toward free truth, but the appearance of truth in the forms of folly.

In this world the most acutely felt outbreak of immanent truth is the appearance of disaster and misery in man's pursuit of happiness. Misfortune is the denial of luck and happiness, appearing as the frustration of happiness in all its forms. The houses of cards, in which people so cowardly lock themselves, will one day all collapse. At the point of death they will all wake up to the awful truth that their lives have been empty, that in spite of all their hard work and meddlesome interference, fate will always keep the world on the course it had mapped for it from the start.

VII: Transcendent Truth

Anyone convinced of the immanent truth of the world of perception, who has understood the inescapable disillusionment of all human endeavour and the inevitability of his karma, will be guided by that conviction in the direction of the reunion of the world with the self – the direction of transcendent truth.

Transcendent truth represents the Kingdom of God in this sad world, self-reflection, forever emanating and reabsorbing itself, the confluence of all fantasies, the *πάντα ρει* of Heraclite. It denies the existence of *fantasies in themselves*, it abolishes desires and fears and also intellectual opinion concerning things which either are desirable or to be feared – as is the case if the intellect is still the servant of a hardened will – or which may be 'objectively true' – that is, when the intellect, living all by itself, has got stuck. In this restricted life it may appear as something unreal, a welcome pretext, satisfying man's need to salve his conscience, and it may also effectively undermine the systems of the restricted life, in this form it is hated by the world and is stubbornly banished, and yet it always returns.

In music and the visual arts, which are understood and felt to stand above life, transcendent truth is accepted, but only in small doses – that is, in accordance with social needs. Therefore the images which these arts produce usually do not represent immanent truth nor moral truth. Almost everything here is either crude titillation, diverting the function of conscience, or an endorsement of society's ideals, temporarily shoring up the flimsy walls of the structure of society's conventions, portraying passions and fantasies which have official approval, so that people can indulge in them with greater confidence, or picturing other passions which might erupt in their culture, just to make people continue to believe that their culture is not too bad after all.

Transcendent truth is not found in art except in the work of a very few, such as Bach and Leonardo. Titillating and anarchical in the worst sense is practically all that is currently considered to be great music or art: the work of Beethoven, Wagner, Rubens, Raphael, and Rembrandt. Examples of *ideal endorsement* are the work of Greig, Michelangelo, and Palestrina, all good church music, as well as the work of Giotto and Memling and all other religious paintings.

Of course, it is impossible to draw the line with absolute precision, in almost any work of art that has stood the test of time there is always some spark of truth, however small; that is the way people like to have truth dished up. However, what one mainly wants from a work of art is *titillation* in times of prosperity, or *ideal endorsement* in times of strife and hardship.

In language, transcendent truth cannot be revealed – even less than immanent truth – without causing an outrage. A clear statement of truth, seriously and emphatically pronounced, is no more acceptable than the manifest performance of miracles. Everyone feels such pronouncements of transcendent truth to be aimed directly at him, that he is more or less told to give up this life of wickedness and folly on pain of hellfire. One cannot gild the pill with stimulating beat and rhyme or melodious sounds, not even if – to avoid bad feeling – one adds that it should not be taken too seriously. Most people have come to think of the Church as something which is not part of real life; its role is confined to the pulpit where the preacher beats about the bush and does not state too precisely what is wrong. Even the work of dead authors – usually taken for a fantasy of times long past and always smugly considered to be somewhat pathological – requires some considerable dilution. Spinoza is a case in point. In his work truth has been watered down so much that it is unrecognizable and everyone can interpret it in the way it suits him; even socialists understand the book to be wholly in support of their practices. For contemporary authors dilution has become obligatory to such an extent that those who really feel truth within themselves will not succeed. Moreover, their personality is a live, clear commentary, an open declaration of the hard truth, and leaves no room for misunderstanding; they show it even if they keep their mouths shut. They will therefore be bitterly resented by most people and be attacked even by the best among them, their admirers, in dutiful defence of mediocrity. Meanwhile they become less vulnerable because of the truth which they uphold; all the suffering inflicted on our Saviour failed to make any impact; even crucifixion did not touch him.

Transcendent truth in language, therefore, has only managed to break through in the work of imitators – those who vaguely understood the word of the prophet and recognized its truth and whose personality has undergone the necessary watering down. They are the ones who are honoured in their circles as wise men or as men of great genius; their appearance gives the message that what they say – in flagrant contradiction with the nature of truth – must not be taken too seriously. People find them pleasant and interesting; the more so because of a certain mystique that surrounds them, because they do not quite understand where such a person gets these ideas, that are so much at variance with his outer appearance and behaviour. Of course, the main concern of these imitators is to keep the prophet away from the circle where they are the stars, that are anxious to preserve the aura of mystery. They will first try to deny the existence of their spiritual father and then disclaim any connection with him. They need not bother, for the relation between the real thing and the watered down version is rather difficult to see anyway.

Writings of transcendent truth which have been preserved are usually the work of an imitator. Their real spiritual father never had the inclination to write; he radiated truth throughout his life, infinitely stronger than he could ever express in words or in writing. He was never able to water down the truth as society demands, neither did he feel inclined to allow the truth, rising so high above the world, to go begging here on earth – allow truth, which transcends language, to go begging in words.

He will also scorn any attempt to make just bits of truth acceptable to his fellow men by appealing to their limitations, their fears and desires, by confronting them with the awful consequences of their thoughts and their actions, or by showing them how their various desires counteract one another and how their fixed ideas contradict one another. He does not want to disturb the self-revenging power of evil, since he knows that removing one desire or error will only make room for another, that man's will is naturally drawn toward passion and folly, and that, deprived of one, he will soon rush toward others.

But imitators seize upon this tactic with great gusto. They claim to be the great liberators who will rid the world of all evil, folly, and injustice, they will be hailed as the benefactors of mankind, but they will leave mankind as miserable as ever it was. They expose the folly of popularly held beliefs but replace them by others, equally stupid, and leave mankind as stupid as it was before.

Knowing that nearly everyone craves the respect of others, wants to be thought of as superior, better than everybody else, able to say, 'Thank thee, Oh Lord, that I am not like any of these people!' and to feel important because of the faith which they – and others do not – profess to, they start up associations of vegetarians and theosophists, and, indeed, even a socialist society of property owners, seemingly unaware of the absurdity of calling oneself a socialist while hanging on to one's capital. They managed to make people less jealous and less greedy – a rather trivial exercise – in the past by maintaining that all their good deeds would be rewarded a thousandfold in the after-life, and do so nowadays by pointing out that a life of love and brotherhood is the ideal state of man and that all who aspire to this ideal do good and are better than others. Sometimes they even maintain that acts of charity and love somehow have the effect of making one's face more beautiful and serene and so will be visible to all with eyes to see.

The pictorial, cosmic system added in the past was one of heaven, angels, the last judgement, the elect, and either eternal happiness or eternal damnation; and, nowadays, as one of cosmic rays, magnetism, somnambulism, reincarnation, and the seven heavens, always on the understanding that such faith is the exclusive reserve of only the best – those who are ready for it.

Every truth, to be more palatable, is adjusted to suit the audience and 'clarifications' are added. When they say 'Do not seek glory by trying to be what you are not', they add that keeping up appearances leads to worry and illness, and that in the end all appearance and pretence will be shown up.

They do not simply say 'Security is mortal's greatest enemy, every penny of your capital is a black mark against you, and saving is a sin, forbidden by the voice within you', but they must add an explanation and say 'Look at the trees, the flowers, and the wild animals; they too live from day to day and yet they do not look any the worse for it'. Instead of spurning human fear, the passion for saving, they play it down and even seem to accept it when they say 'Capital and property are a barrier on the road to happiness, because hard necessity – hunger and cold – is the only source of pure growth and strength of character, without which there can be no physical health either'. The people will nod and agree, hail them as great and wise men, and then go on saving and living their lives of abuse and exploitation.

They do not simply say 'You should not wear any clothes, because they are a cover of fear, pride and vanity', but in their explanatory comment they refer to the important role that skin-breathing plays in the human metabolism and to the salutary effect of exposing one's skin to the fresh air; they turn themselves in to experts and reformers of hygiene. The idiots listening to them take to airbaths and, when somebody discovers the beneficial effect of sunrays, they take to lightbaths and sunbaths, and finally to duskbaths, nightbaths, moonbaths, starbaths, forestbaths and meadowbaths as soon as somebody proclaims them to be healthy. But all this leaves the people just as unhealthy as before, because their evil nature makes them sin against their health in yet another way,

They do not just say 'Pray and work!', but must add that praying is a kind of recapitulation, a concentration of mind which gives a better view of life's path and helps one to follow that path refreshed and steadfast, a guard against illusion and error. They do not simply say 'You should live naked in the world of nature, leave nature undisturbed, and you should not work', but they add 'You are worried about losing body heat, but should know that our ancestors were naked and they lived at a time when our climate was certainly not warmer than it is now, that some of the ancient tribes walked around practically naked and did so at a temperature of forty degrees below zero. You are worried that nature is not rich enough to feed you and that you will starve unless you work the soil, but remember that whatever nature has brought forth will be maintained by her as long as needed; remember also that Catharina of Siena did not eat at all. Finally you are worried that you might be torn apart by wild beasts and be deprived of your precious life, but you should know that no wild animal will attack a truly good human being because in his looks there is something that the glazed eyes of those who have not got it cannot see, but something that wild animals will recognize. Only when people began to live a life of greed did the need arise for so-called heroes, to fight wild animals.'

They do not simply say 'All forms of transport are evil', but they point to the ill-effect of the smoke of trains on one's health, to the damage to the nervous system caused by the electromagnetic field of electric trams and the inevitable disharmony in the human body which must follow displacement of the field of force without the appropriate use of muscular power. And when they say 'All cultivation is evil; the abuse of nature and its forces is just as immoral as the abuse of human beings and animals', they must point, in addition, to all the degeneration, illness and misery that man's cultivation has brought about. Indeed, they sometimes even express their willingness to enter into a debate with the fools, their fellow human beings! Contrary to their own expressed beliefs, which reject the will of others, they accept this will by the very act of entering into debate, and make it equal to their own will.

In order to make a truth more palatable they will not hesitate to base it on a definite endorsement and proclamation of some fashionable piece of folly. For example, economic reformers base the fact of social injustice and the need for a better, higher form of life on the foolish arguments of fear and ambition, as if the 'higher' can only live on a full stomach – *primum vivere deinde philosophare*, they cry – as if evil, in all cases, can be avoided by rational argument and by action. Often, while explaining their moral disapproval and trying to console their audiences, they tacitly retract their original stands. They do not simply say 'Abandon the illusion of the constancy of matter, the self is sufficient and can create all without any limitation', but they must add explanations and hypotheses on the nature of matter and so, equally foolishly, introduce the constancy of other things like electrons.

Their advice, 'Rid yourself of your intellect, that gift of the devil', is qualified by some added remark which in fact endorses the view of the intellect they had just condemned – for example, 'The structure of nature is so infinitely subtle and complex that your intellect will never fully grasp it and therefore will never give you the stability you are seeking'. For those, however, who mange to relinquish the intellect, the world is anything but subtle or complex; it appears to be subtle only to an intellect which struggles laboriously and sees no end to its struggle.

The role of preachers therefore is no more than that of a guide, helping along without any power the self-correction, the self-development of the life of desire on this earth, where each form of folly is only a temporary craze which soon exhausts itself and is then discarded by the preachers to make room for other follies. Yet people still behave as if the end justifies the means; they foolishly see something desirable and go after it, using means which are themselves felt to be rather unpleasant. For example, they find the whole business of breeding, feeding, and milking cows rather revolting, and yet they expect nothing but benefit from it; they go on consuming meat and milk until they discover that it damages their health. They dislike the planting and pruning, and the use of manure in horticulture, and yet they expect nothing but benefit from it; they go on eating fruit and vegetables grown that way, until again one day they discover the harmful effects. They dislike spinning and weaving, and yet they go on using curtains and clothes until that too is shown to be harmful. They resent all hard work and the rat race, and yet they expect all kinds of benefits to flow from this 'culture'; even those who become disillusioned and give it all up still go on, elsewhere, pursuing goals and working hard, because ambition and hard work are part of their nature. History shows that the form and shape of the human prison may change, but its walls are never demolished.

Lyrical poetry is a form of *wordcraft* and has therefore nothing to do with transcendental truth. It dreams of all kinds of states of mind which flourish in the madness of our culture; according to whether it is more or less intellectual, it either supports the ideals of the reader or simply titillates.

It sings of love and sadness, of passion and despair, moons and daisies as they all appear in this imprisoned life. The reader hears the echo of his own fantasies and has the pleasant feeling of being supported, of greater self-confidence and contentedness, all of which he badly needs. Most to his taste is the added infusion of a little *Weltschmerz*. This mixture seems to give him comfort and a promise of *balance*, grown from *Weltschmerz* with the help of some overdeveloped emotions; it helps him escape from his own feelings of dissatisfaction in endless dreaming and gives a temporary release from pain.

Truth and poetry, like any other merchandise, are falsified: indeed they are hardly ever found in a pure, unadulterated form. Most philosophers and moralists believe as little in what they write as the manufacturers of baby foods and meat extracts believe in their own products; neither do they act with more good faith than those who lead spiritualist séances, and very few poets have themselves experienced the happiness they describe. The critical sense of our corrupted instincts is somewhat warped, it does not detect fakes: *mundus vult decipi*. Priests do not believe in what they preach to the masses; the leaders of political parties deceive the people deliberately, using words they don't really understand. Most poets, painters, and other artists have abrogated this role to themselves out of weakness or laziness, knowing themselves to be poorly placed for any other role in the social industries, and the uncritical public has come to accept their place and their fake products simply because they cannot do it themselves.

Sometimes only the accompaniment of transcendent truth may be heard in life. Truth itself is absent, remains outside this limited life and therefore outside the domain of communication and mutual understanding; its expression seems to be completely removed from life of which it is a part. Returning to his humble, earthly duties the 'seer' will steadfastly believe in the sudden flashes of imagination received in self-reflection as the accompaniment of higher wisdom, recalling the echo of the guiding voice of self-reflection.

These images are the harmonious results of an attention to the self and of work in this world. They are not an expression of an attention to this world. He who lives in self-reflection, in freedom from fear, desire, and knowledge, who does not see nor follow any direction in this world, who only does what he is made to do and in this way guards himself against irreversible actions which only aggravate his karma, who is not affected by outside influences and stands aloof from what happens outside, who does not grow but quietly maintains his position and at the same time feels free to remain motionless outside the world where he has escaped from his karma, from misery, from growing old, from decay and death – such a man will see even the flashes of imagination of others as accompanying the truth in his own life, moving high above the world and detached from the forms of this world.

Those imprisoned in life call it mysticism. They think it obscure, but in truth it is light; it is only darkness to those who are in darkness themselves. The fantasies of mysticism are locked in forms which come closest to the humble but sacred task it must fulfil in this world; they therefore do not so readily appear in music or in the visual arts and are usually expressed in words, which are closest to the human curse, the intellect. Mysticism simply denies that there is anything positive to be found in this limited life; it reflects the infinite emanation and reabsorption of the self in strange imagery and sounds. Only those who know the melody can understand the accompaniment, and they will recognize it even if it is a strange accompaniment on a strange instrument. Such are, for example, the images of ancient and medieval mystics; they were taken from the perceptional world but seem strange to the modern reader who has not learned to see such a pantheistic world in the surroundings where his duties lie. But that is no reason why he should not understand them.

Sometimes the sounds of truth and of mysticism do not follow the rules of the intellect, but they can be understood by the intellect. Such were the words of those who had been able to turn into themselves. Living their imprisoned lives, they treated this experience as something independent, something outside themselves; they used it to strengthen themselves through words expressing insight and moral sense within the system of this limited life to which their attention had reverted. These writers may be called semi-mystics. Their work is a source of irritation to those who understand because it brings super-natural truth too far down to earth. For those who do not understand, their work is extremely dangerous because it leads the wanderers searching for security and certainty to all kinds of extremes. At the bottom of almost all forms of religious extremism and sectarianism there is always some semi-mystical pronouncement of truth.

The Church was therefore quite right in condemning the heresies of Eckhart, Huss, Luther, and Calvin who brought down to earth what should really have been left high above the earth. These men could not remain steadfast, unaffected by the instability of this earth and accept it as God's will; they were moved by their own instability, allowed it to affect their own wills and tried to bring greater stability through their own wills. A much purer form of mysticism is found in the writings of the ancient Indians and Chinese and some of the Church fathers, as well as in the work of Jakob Boehme; their writing is kept well above the level of practical everyday life and is beyond practical understanding. Great intellects may sometimes manage to make a photographic copy of these works within their limited brains and that copy may then seem to hold great truths; however, the essence of the original is completely lost. Since it is all beyond the wits of the hoi polloi, the poor souls suffer no harm from this little game – for it is no more than a little game when so-called philosophers start giving rational explanations of God, Trinity, the

Immaculate Conception, and the Brahma-Vishnu-Shiva; nor is biblical exegesis more than a little game.

To the intellect, and in the context of this limited life, pure mysticism is totally meaningless; it does not stir the conscience of evildoers and leaves the great and powerful in peace, and they in turn do not bother themselves much with this 'harmless curiosity'.

And since mysticism remains outside the sphere of reason there is not much that can be said of it in rational terms except perhaps in the negative.

Mysticism cannot be learned; it can only be recognized. Anyone with talent can write the truth, and talent can be found even in imprisoned minds; moreover, anyone of sound mind and body can understand the truth. But writing and recognizing pure mysticism requires a freedom of soul, which cannot be acquired by earthly means but can only be granted by divine grace.

Mysticism is quite different from occultism, or rather it is its complete opposite: mysticism denies all knowledge whereas occultism follows the human thirst for knowledge to its extreme; occultism is not concerned with morality, while mystical wisdom and high moral sense are inseparable companions. Nowhere in mysticism is there a thread or a correct sequence. Every sentence stands on its own; it does not require another sentence to proceed or to follow it, as suits the accompaniment of what is timeless. It treats the questions posed by metaphysics, such as immortality, free will, the meaning of art and religion, and the foundations of morality, as riddles hatched by the intellect; in doing so it removes all mystery and yet shows the impossibility of solving such questions by reasoning.

To the intellect mysticism sounds incoherent, oracular, and sometimes even bombastic – something that cannot stand the test of criticism, all-bristling with contradictions. Here are some mystical sounds which illustrate and transcend what was said in the previous chapters.

> The first property of nature is desire; it is like a magnet, a drawing power of the will, which wants to be something but has nothing of its own to build on; it therefore acts through an attractive force and fastens itself on to a *something*, and yet it is nothing but an acutely felt magnetic hunger, a bitterness, hardness, and cold.
>
> The second property originates in the first, it is the act of drawing, of movement in this acute state. The magnet (the attraction) makes hard, and hardness is again broken by movement, and therefore there is an everlasting battle within the self; this causes bitter pain, stinging its sensitivity, which could not exist without acuteness and movement.
>
> The third property is fear or anxiety, a form of will induced by the attraction toward nature and the ego. This movement is like a turning wheel, because desire pulls toward itself while movement repels. In such a state of anxiety the will can neither move inward nor outward, and yet it is pulled inward and outward. This state of fear is the real foundation of hell in as much as it is not – unlike God's being through eternity – absorbed and raised in the freedom of light. (Boehme, II, p. 57)

> Creatures should remain humble and obedient to God and not try to raise themselves higher. They are not God's equals: God wants the company of children, not of masters: He is the Lord and no one else. (Ibid., p. 65)

Ever since God spoke the word of creation, the wheel of eternal *essences without being* turned in wonder. However, when he moved will into fiat, it turned into being. And then time began, time which in eternity did not exist before. (Ibid., p. 66)

As the revelation of the eternal conflicts with that of the external, earthly, and fallen natural world so do the spirits of the dark world conflict with those of the holy world – in particular in the case of human beings – manifest in a battle of good and evil. In this way God has set one against the other, so that his majesty will be revealed both in his love and his wrath. (Ibid., p.78)

The angels are our servants and our guardians that we may be Christians and not animals. (Ibid., p. 79)

There must be a struggle until the dark, hard, and closed kernel cracks open and the heavenly spark sets it alight, so that from it – as Christ says – will grow a noble lily as from a divine mustard seed. One must pray seriously and in great humility, sometimes even spite one's intellect and be a clown, appear to be foolish, before Christ will take form in this new incarnation. (Boehme, I, p. 83)

It is quite possible that a poor dead sinner mends his ways, when he stops and moves away from visual images and listens to the Lord's voice within him.

But the impenitent, the embittered will not listen to the Lord's voice within, he only says: I want words! printed words! To him only the written word matters, he plays with these words and abuses them and he prides himself in doing so, but he ignores the living Word that has spoken them and he does not want to listen. If, however, he sees the light and repents, he will be dead to the old printed words and be made alive by the Spirit behind them. (Boehme, III, p. 215)

For even though a wise man strive, O son of Kunti, the forward senses carry away perforce his mind.

Holding all these in check let him sit, controlled, intent on me: for he whose senses are restrained possesses steadfast wisdom.

When a man ponders on the things of sense, springs up attachment to them; of attachment is born desire, of desire is born wrath.

From wrath there comes delusion, and from delusion a wandering of memory, from memory wrecked the ruin of reason; with reason's ruin the man is lost.

But he who approaches the things of sense with sense from love and hate disjoined and under self's control, with governed self, comes to serenity.

For him serenity begets the loss of every pain; for soon his reason becomes steadfast whose mind is serene.

There is no judgement in the uncontrolled, and in the uncontrolled is no reflection, the unreflecting man can know no peace; he that has no peace – whence has he pleasure? (*Bhagavad Gita*, II, 60–66)

A man should not rejoice at gaining what he loves, nor grieve at gaining what he does not love, steadfast in judgment, undeluded, knowing Brahman, in Brahman, abiding.

He who, with self detached from contacts without, finds happiness in self, enjoys imperishable happiness, his self controlled by contemplating Brahman.

For the joys that are born of contact are surely wombs of pain; they have beginning and end, O son of Kunti; not in these does the wise man rejoice.

He who can bear even here, before he finds deliverance from the body, the impulse that desire and wrath beget – he is controlled, he is the happy man.

He who has joy within, pleasure within, and light within, the ascetic, becomes Brahman, and reaches the calm of Brahman. (*Bhagavad Gita*, V, 20–24)

When thought is curbed by practice of control, and comes to quiet, when he sees self by self and is with self content.

When he knows that utmost pleasure which can be grasped by reason but is not reached by sense, and when he stands and swerved not from the truth;

Than which, when gained, he holds no other gain more excellent; wherein he stands, and not shaken even by grievous pain;

This disunion from union with pain, he should know, is called union by control; this control must he practice with firm resolve and undespairing heart. (*Bhagavad Gita*, VI, 20–23)[3]

VIII: The Freed Life

In this world, the self and transcendent truth are also reflected in the everyday life of those who are free, those who expiate their old inevitable karma without creating a new one, those who humbly accept their incarceration and never try to break out violently but who, on the other hand, never hesitate to leave as soon as the gate of liberation is opened, those who feel and accept their bodies as straitjackets until the day when it pleases God to free them and take them to Him. In the eyes of their fellow men their influence in this world is insignificant, and yet they are the ones who carry out the ordinances of fate, the appearance of chance or accident which spites the laws of causality, turns the course of life and takes its revenge. Their well-intended action, however, will not turn the course of events; they cannot disturb the self-revenge of evil and therefore they refrain from preaching the truth. And yet, in spite of their humility and reticence the truth will show in their personalities; it will cause uneasiness and resentment and sting others into acts of violence against them. From the evil and suffering inflicted on them will spring the various religious movements all over the world, apparently as something positive and new, but in fact no more than a negative revenge on the old.

Their lives show a disregard for pleasure, property, honour and even of work – except the tasks immediately before them. They have not set themselves any targets either for the immediate future or for the whole of their lives; neither do they seek contact with their fellow human beings. Societies are for them the lowest dregs in the thick fluid of mankind.

A man therefore will move toward absolute solitude. Not so for a woman: her life will always be a searching and a giving of herself in human – that is, male – company. His unchangeable karma, through which and from which his life is directed, is the environment in which he has been placed as well as his reaction to this environment in accordance with his fateful intellect. Hers is a hankering after that which is male as it lives in her beloved; this yearning for what is human outside herself and absolutely separated from herself causes her life slowly to move toward purity. She will ignore her surroundings, life's conditions, and even her own human faculties. Her life will remain directed towards her beloved and therefore stay in this world; she feels she cannot escape as long as she is not man herself. But his path leads away from this world; as soon as he becomes aware of his manhood he moves along this path toward atonement and ultimate elimination.

To start with, he will make his intellect adapt and conform meekly and zealously to the habits and ideals of society, and he will listen carefully and wait patiently for the revelation of inner contradictions of that intellect. This revelation is not forced by him and therefore it will only appear when he reaches the ultimate consequences of philosophy and gets stuck as in the vertex of a cone. At that moment the illusion of the world evaporates and the self is revealed. From then onward science and reasoning will disappear from his life, recognized as the mere products of arbitrary limitation. He now only lives in the present moment, happily accepting his condition and his environment, reacting to it with equanimity and carefully waiting for any opportunity to escape from its oppressive force. After his first little escape he feels no longer at home in the company of his fellow men, and they get irritated by his eccentricity which follows from his newfound freedom. He in turn begins to resent all his links with society; he is forced to exercise extreme care in human company and faces the difficult task of reacting to circumstances without being susceptible to the suggestions implied by these circumstances.

The temptation to give up his new freedom and fall back into the old routine would indeed be great were it not for the all-embracing power and wisdom which flow to him from his self and which, together with mysticism, guides him securely on his path and, through immanent and transcendent truth, keep life's blood warm and flowing when there is danger of curdling and freezing and therefore of its ultimate death.

These disturbing influences can only help his patient move away from human society. His needs, the burden of his body, will steadily grow fewer and those that do remain will increasingly be met by his own hard work rather than by sponging on others; in the end, his dislike of labour will lead him to eliminate all needs. In this way there will be a steady cleansing of his environment without any interference on his part; the intellectual mists – another burden he has suffered with equanimity – will vanish. His intellect life will be like a forest path: from afar the end will seem dark, but as he walks it will gradually become clearer and clearer.

He will go on and reach a state of ever greater solitude, poverty and immobility: the last that society will see of him is when he disappears – a hermit seeking the barren heath over lush but dull vegetation, seeking the night rather than the insipid light of day. Often he will bathe in the ocean. He knows that he is destined for even greater poverty, and lives in the belief that:

> Poor are those who are not satisfied with all that God has ever created. Poor are those who do not want anything, do not know anything, and do not have anything.
>
> If man is at the stage that there is still some place for God to work, we say: as long as there is such a place in man, he is not poor in the deepest sense, for God is not at one with his work. Man should have a place where God can still work: it is poverty of spirit when a man is so empty of God and his work that God when he wants to work in man's soul is himself the place where he works and that he does willingly.
>
> There man is what he was, and there he does not grow nor decline because there he is an immovable origin which moves all things.

Here God does not find in him any place since man because of his poverty achieves what he has always been and will forever be. And here God is one in spirit; this is the deepest poverty we can find. (Meister Eckhart)

What little he does is done reversibly, that is, it can be undone; he does not care whether his work succeeds or not. Since all his actions are reversible he can let himself go in good as well as in evil. Sometimes he will do evil, sometimes he may seem to tighten his earthly shackles: the ways of the self are inscrutable. Maybe he will return to the old life and stay there without regret, appear to be driven by passion, beyond hope and heading for hell. But it does not disturb him; he stands outside the world where he has no obligation; he cannot sin anymore, he cannot do anything there anymore, he has been dead for a long time, his attention moves in higher spheres and 'Apostasy is permitted as long as the heart is pure' (Flaubert).

Only the death of aversion from God leads to true tranquillity. He who has forsaken his self and surrenders himself wholly to God in mind, taste, desire and will, he is dead to this earthly world; there is a split within him: the aversion, self-centred, keeps stirring in the self unto death, but the surrender of will lives on in the death of Christ and will remain in his resurrection in God. And even if his inborn passions lead to sin – and they can do nothing but lead to sin – the surrendered will does not share in sin, because it is dead to passion and sin and lives through Christ in God. It lives in the land of the living, while the self-centred will lives in the land of the dead, an ever-dying. (Boehme, III, p. 263)

Now that man whose delight is but in self, whose pleasure is in self, whose satisfaction is in self alone, has no work that he must do;

For him there is no purpose here in work done or left undone, and he has no reliance on any being for any end.

Therefore without attachment ever perform the work that thou must do; for if without attachment a man works, he gains the highest.

Entirely by the strands of nature are works done; he whose self is deluded by the I thinks 'I am the doer'.

But he who knows the truth about the distributions of strands and works, O thou strong of arm, thinks 'Strands abide in strands' and so escapes attachment. (*Bhagavad Gita*, III, 17–19, 27–28)

Following the path of practice, his self refined, his self subdued, his senses conquered, the pure self becomes the self of every being, although he works, yet he is not defiled.

'I do not work at all' thinks he whose way is practice, who knows the truth, although he see, hear, touch, smell, eat, walk, sleep, or breathe.

Speak, let fall, lay hold, open or close his eyes; remembering ever that the senses abide in the things of sense.

He who lays works on Brahman, abandoning attachment, and so works, is not smeared by sin, as a lotus leaf is not smeared by water.

With body, mind and intellect, and sense alone, ascetics do work abandoning attachment, to purify their selves.

He whose way is practice abandons fruit of work, and wins to final peace; he who shuns practice, and is attached to fruit by the prompting of desire, is bound. (*Bhagavad Gita*, V, 7–12)

But the free man – whether he continues his escape through every possible opening or returns to the old life – does not touch the walls of life nor does he feel trapped in them.

That is why his beauty shows through these walls. Only visible to his equals, his beauty shines through all that binds him: his house, his clothes, his country, and his body as his 'idea', the karma which burdened him from birth but above which he has raised himself. This beauty is free from the world, free from decay and it is imperishable; it is the beauty of freedom, visible through chains, since freedom at a time of tribulation is always enchained.

As long as he has not moved out of society there will always be women whose lives flow toward his. If they are driven by passion or seek support in his weak manliness, their lives will not penetrate into his life because they do not see him. But if they live pure lives – that is, live in him without lives of their own – then they will see him and their lives will flow and merge with his. The lack of any life of their own does not allow them to break themselves loose from society; it is therefore through his women that he will maintain his last links with society and through some very fine threads remain rooted in society. The temptation that radiates from them unconsciously will appear to him as the strange and tempting radiance of the flowers of his dreams, which vanishes when touched and which kills the guilty, a radiance which respects the lives of those who pass it by with reverence and admiration, a radiance which is immaterial and therefore out of reach of any attention imprisoned in plurality and materiality, but a radiance which without any resistance will penetrate into the soul which is pure and does not need the support of matter.

The women who swarm around him, driven by passion, are like vampires but like the predators of his solitude they can cause him no harm. In the end he is no longer seen among his fellow men because 'None of all that God has created satisfies him anymore'. If he is dead, he has escaped from his karma and therefore from existence and limitation. However, escape from his karma alone does not necessarily mean that he must die nor that he must live in a country with unknown horizons and strange fellow creatures. He will therefore live alone and naked on a desert island, not too big, and not too small – one that he can easily oversee. Is it to be found in this world? No, because there is no longer any need for an earth and therefore earth no longer exists.

The sea is calm, the horizon sharply set. The needs of his body, which he already knows to be non-existent, have now also vanished physically. There is therefore no longer any reason why his body should die. He does not eat anymore; he surveys his island all around him – a fox, a few rabbits rustle away and some birds perch quietly in the branches. He sits down on the beach and watches the horizon. There is a soft fall of rain; in the sky behind him the moon, and over the sea a pale shimmer. The birds, also behind him, watch him silently, wondering but paralysed. It is all frozen in time: so it has always been and will forever be.

That was mysticism. It is Flaubert's immanent truth in the cloak of fantasy of his *Gymnosophiste*:

> At the edge of the forest there stands something that looks like a woodpile, something rather strange: it is a man, completely naked and covered in cow dung, more emaciated

than a mummy; his joints are like knots at the end of bones, which look like sticks. There are shells over his ears, his face is thin and long, his nose like the beak of a vulture. His left arm points in the air, paralysed, stiff like a wooden stake; he has stood there so long that birds have nested in his hair. On the four corners of the woodpile there are four fires burning. The sun shines right in his face, he watches it with open eyes.

[And while the flames surround him he says:] 'I am like a rhinoceros, I have buried myself in solitude, I lived in the tree behind me and fed myself on flowers and fruit, so carefully observing the rules that even a dog could not see me eat.

Since existence arises from corruption, corruption from desire, desire from sensation, and sensation from contact, I shunned all action, all contact, not making any more movement than the stone figure in the graveyard, breathing through my nostrils, my gaze fixed upon my nose. Seeing the universal ether in my mind, the world in my limbs, the moon in my heart, I became aware of the essence of the Great Spirit from which keep bursting forth the beginnings of life, like the sparks from a fire.

At last I found the Supreme Spirit in all beings, and all beings in the Supreme Spirit; and I was able to make my soul enter into it, my soul into which I had brought all my senses.

I gain my knowledge directly from heaven, like the tchataka bird which only uses the rain to quench its thirst. Only in this way do I know things, things themselves do no longer exist.

For me there is now no hope, no anguish, no happiness, no virtue, neither day nor night, neither you nor me: absolutely nothing.

This awful deprivation has made me stronger than any power; one movement of thought can kill a hundred princes, dethrone gods, turn the whole world upside down.

...I have taken a dislike to form, to perception, and even to knowledge itself because thoughts do not survive the transitory facts which brought them into being; like everything else, mind is an illusion.

All that has been created will perish and all that is dead will revive; beings that have actually disappeared will dwell in a womb not yet formed and they will return to this world to share the suffering of other creatures.

But having turned through an infinite multitude of existences in the form of gods, of men and animals, I now give up all travel; I do not want any more tiredness! I abandon the squalid tavern of my body, built of flesh, coloured by blood, covered in a hideous skin, full of dirt. Instead I shall finally go to sleep in the deepest sleep of the absolute: in annihilation.'

You may think it strange that the way of life chosen by the greatest, the wisest of men requires so few special skills and powers, that such total surrender, such laziness, is too easy. But, remember, they have broken right from the start with the habit of aspiring to the difficult and the artful, of making the best of one's talents; for a long time there has been nothing in this world that they admired and worshipped; they know that there is nothing here worthy of worship or admiration. They look upon their talents as temptations, luring them toward action in this world (which again promotes itself as the real movement of the world), temptations which remain powerless, frustrated by their insight of the 'grandeur and decadence' of all worldly greatness and worldly shows of strength. From the moment that the inner contradiction of the intellect was revealed to them they shunned and ignored their own talents. They do not wish to know the truths of the world nor learn the art of prediction, knowing full well – even if they can see and feel through past and present – that this has nothing to do with the world of outer appearances nor with language, and certainly cannot be expressed in language.

They can do anything that comes their way, but rather than perform tricks they work miracles.

As to the free woman, her life's path moves through the densest growth of society and there it will end, even if her real self remains outside, existing as it does outside time and space. She will seek nor recognize stability, she trusts no one nor will she be faithful to anybody or anything. But the barrier wall built around her by her karma, her fate, is to worship the highest in man that she is capable of grasping, and to adapt her own external life to the service of his free and full development. In doing so she will not be bound by any law or convention of society.

Uninhibited, she will spread crime and cruelty along her path; she will show the same lack of restraint in guarding herself and her own life – itself of little importance – in the service of her beloved, not afraid to sacrifice human lives in the process and not ashamed to lose her honour. He is the only one she tries to care for, but he does not know her; and if his life would touch hers, she would choose to serve him in the most menial and degrading of tasks, knowing full well that nothing in this world is unworthy of woman, who has no soul and bears the guilt of man's fall.

The barrier around her will open up when her beloved falls away from his karma for good, revealing the contradictions of the karma that abandons itself, revealing also the illusoriness of her ideal. It may also happen that a higher male karma is revealed to her, that a great light will shine through the old barrier walls, which will appear to be nothing, the back wall eliminating the front. A new deeper and lighter wall then becomes visible behind the old one and demands her full attention. Her heart now belongs to her new beloved, the old one deceived by her fake loyalty – maybe to his salvation.

For the free man, life is a full but humble employment of his intellect reacting to his environment; for woman it is love, that is her inexorable karma…until the last, the faintest and lightest of walls has come down and she falls into an empty space: the mirage of male stability outside herself, the only basis of her femininity, has faded away; her life no longer has any purpose and after this last revelation it has crumbled into nothing.

Having served many, one after another and with greater purity at every stage, she meets her last friend, in whom manliness has raised itself and as the highest male principle returned to his father, the last friend whom she can only serve by leaving. And she goes away and lies down awaiting death.

IX: Economics

There is one more evil which the free life will carefully avoid as long as its links with society remain, and that is 'economics'. It is governed by the absolute conviction that folly and injustice are an essential part of society; indeed, if society were better, if it were governed by love and brotherhood, there would be no grounds for its existence; it simply would not exist. There is no temptation to look at it more intelligently, examine the manner and the rules by which misery and injustice operate. To the free man, the world is no more than the brute force and restraint which it uses against him, the guilt with which it burdens him and which haunts him. This doom is all he thinks about: he thinks of nothing but escape, of freeing himself slowly, avoiding falling down the ravine of temptation, not pursuing what is desirable and not trying to make up for what is regrettable. For if one considers something to be desirable or regrettable, one sees it as something outside oneself, as part of a world which is independent and permanent, as part of an inalienable property which can be nurtured, cared for, improved, and made to grow, like flowers or chickens. Trying to exert outside influence for the sake of a better world or one's own power is vanity, blind folly, and lust for power.

The free man rather looks on his fellow men as burdensome hallucinations, luring him away from the right path and trying to make him join their ways because they cannot tolerate his freedom. The free man will carefully avoid them.

Messing about with society, trying to change it, is something we will leave to 'idiots with ambition', knowing full well that there will always be such idiots and unworried about the lack of any such people. For if there were no people chasing the mirage of a world to be governed by them, the world would be perfect; there would be no need for government and no need for social work. If the world were perfect, it simply would not exist. The self-correction of the social lie is guided by the hand of truth, but truth dressed in the robes of lies, and therefore it does not stop at cleansing and breaking old injustice – old folly: from man's penchant for folly and greed will grow new folly and new injustice.

Look and see how the theories which help to undermine one worn-out social structure of injustice always carry in their own positive concepts the germ of yet another evil, just as deadly as the old one.

They talk about the 'rights of man', as if human beings bring with them into the world, rights, and worse, duties as a punishment for being born.

They talk about 'work', the necessity of it, and the happiness it brings, as if human labour were more than a blind convulsion of fear of what in fact is not evil at all, and of desire for what can only bring misery – this wretched 'work', through which the human swarm of insects has pushed back and eaten away Mother Nature, who gave them life and kept them in balance! They will all end up lonely, in utter misery, without balance and support! In order to extend their miserable lives they use the services of such hellish forces as fire and cohesion. Work, which uses its hellish powers to create wealth, which turns into sensuality, which makes the human species expand even more, makes it more miserable and more dependent on the services of hell; work, in which man's sin has turned the world into a necessary instrument of fear and greed, and made it a place of sorrow, folly, and misery!

They talk about 'the poor and repressed' and about 'repressed classes' as if anybody in this world were born into a state of repression without deserving it! As if anybody could suffer social deprivation without having bowed down out of fear! Is not the world a garden of misery, kept alive by inner contradiction, a place where everyone, after a helpless struggle and total defeat of his ambitions, receives his reward according to his work and his guilt! And will this not remain so forever, and will people not fight their punishment only to protract it even further?

They talk of human talent and the joy of life, which the poor repressed classes have no chance of developing or enjoying. But there is no joy of life; it is only an object of desire; life is joyless! And as to talents, they only tempt and divert attention away from life's path; by ignoring one's talents one usually saves oneself a lot of folly and misery. In the rare case that the deployment of talent helps a man to keep to the free path of life, toward atonement and redemption from his karma, one will find that he has maintained the bond with the self, and that in this case he would have succeeded even if he had been born in the most depressed of classes.

They also talk of 'justice' and, in their childless and arrogant expectation of the future, they cry that one day justice at last will rule this world. But is not justice merely a way of keeping human society in a kind of frozen state, expressing their separation without independence? Has not mankind united itself in justice out of a common fear of the uncertain and fear of one another, not realizing that in doing so they have done nothing but shift the area of uncertainty, moved the battleground from where they openly murdered one another by every means available? Having justice on their side, the battle now becomes fiercer and more ugly than at the time when there was no 'justice'. It started with the duel, which lacked spontaneity both in the opening of the fight and the choice of weapon. Nowadays not only have fistfights and duels been abolished, but we must even pay our bills and we are not even allowed to falsify our signatures! And that all in order to protect the devious power of money and those parasitic state institutions. The venue of permissible fight has now moved to a most unpleasant backyard; that is where the centre of gravity now lies, where people now murder and swindle one another. Mind you, those who do not like fighting there and simply choose to ignore justice and the law do so anyway: fighting against justice is much more provoking and also more dignified than with justice on one's side. Those who know no fear will also find justice to be quite an easy match. Don't worry, ninety per cent of all murders are never solved; moreover, it is probably the best solution for those who were murdered, and they deserved to be murdered.

Economists and political leaders love to speak of some 'future state of all people working consciously together'. Such a state would only be possible for people without fear and greed; but people like that would never work and therefore such a world is an impossibility. Among all the people crowding together in common greed, swallowing all this wonderful drivel, there is not a single person who does not know full well that he himself would never satisfy the demands this future state would make of its citizens.

The masses are so blinded by greed that they do not even notice that their own leaders live in ill-gained luxury, in cosy isolation and at their expense. Indeed, they do not see that in their political party there is an even more degrading regime and greater repression of personal fulfilment than that suffered by the most repressed subjects of the state which it opposes. Socialist workers are the slaves of their leaders more than of their industrial masters. In the past people fought their lords, fooled by the promise of the liberation of the nation, national freedom; nowadays it is the promise of freedom for the working classes. But the people themselves will never be liberated; they will continue to be oppressed and exploited out of greed and for the sake of the mad fantasies and ideals of some individuals. Oppression and exploitation are only shifted elsewhere by the apostles of the new truth; injustice will live on, generously fed by fear, greed and folly, and in its prevailing form supported by some new aspect of 'justice'.

No, the world can never be reformed so as to bring good to man. Social conditions will remain wretched and life for every individual will remain a misery, only aggravated by hope for advancement and a better future. Only complete surrender and total resignation would end this misery.

Look at this world, full of wretched people who imagine that they have possessions, worried that they might lose them and ever toiling in the hope of acquiring more. Look at all these people, striving after luxury and wealth, those whose riches are secured, whose stocks and shares are safely deposited and who now nurture an insatiable appetite for knowledge, power, health, glory, and pleasure.

Only he who recognizes that he has nothing, that he cannot possess anything, that security is unattainable, only he who completely resigns himself and sacrifices all, who gives everything, who does not know anything, who does not want anything and does not want to know anything, who abandons and neglects all, he will receive all: the world of freedom is opened to him, the world of painless contemplation and – of nothing.

Notes

1 *Life, Art and Mysticism* clarifies many points in Brouwer's philosophy of mathematics: his stance against logicism, the rooting of mathematics in language, and Brouwer's notion of intuition. Readers who wish to investigate further should study the following texts by the translator: *Brouwer's Intuitionism* (1990), North-Holland (Elsevier), Amsterdam; *From Brouwer to Hilbert* (co-authored with Paolo Mancosu) (1998), OUP, Oxford; 'Introduction' to 'L.E.J. Brouwer, *Life, Art and Mysticism*' (1996) *Notre Dame Journal of Formal Logic*, **37** (3).

2 First published in Dutch, as a booklet, in the Netherlands. This translation is reprinted from the *Notre Dame Journal of Formal Logic*, **37** (3) Summer 1996, pp. 381–429. It appears in *Post-Analytic* Tractatus, by kind permission of the *Notre Dame Journal of Formal Logic* and the translator.

3 In the case of the Bhagavad Gita, Brouwer simply quotes 'chapter' and 'verse' (e.g. II, 60–66) and quotes from a German translation, in German, without giving any source. The translator has used an English translation by D.P. Hill, OUP, 1928.

- Is this the kind of chatter W (TLP) would've objected to?

Chapter 2

Logic and Ethics as the Limits of the World

Anthony Rudd

In this paper I want to focus on the notion of 'showing' or 'making manifest' in the *Tractatus*. I shall argue that to take it with the seriousness it deserves, we shall need to adopt a particular reading of the work as a whole – one that stresses, more than is commonly done, the influence of Schopenhauer on it. That Schopenhauer was a significant influence on Wittgenstein's early thinking is now generally recognized, but the nature of that influence is still not always well understood. I think that, at the time of the *Tractatus*, Wittgenstein accepted a great deal of Schopenhauer's metaphysics, although, according to his own restrictive theory of meaning, the deep metaphysical truths that he saw in Schopenhauer's work could not be stated in literally meaningful language. First, then, I shall give a brief exposition of the themes in Schopenhauer that I think influenced the *Tractatus*.

I

Schopenhauer's philosophy is characteristic of German Romanticism and Idealism in its emphasis on the notion of expression. The Romantics reacted against the 'scientific' idea of the world as a set of distinct, but ultimately homogenous, entities linked to one another by external relations of mechanistic causality. Instead they revived in a way the premodern idea that things were linked to one another by what they expressed. The most obvious case where this notion seems to have application is the mind–body problem. Instead of supposing that mind and body were linked by quasi-mechanistic causality, or that mind could somehow be reduced to body, one may see the body as what expresses the mind. There is nothing particularly mysterious or esoteric about this idea; it can be illustrated from any ordinary social interaction. Normally we are able to see another person's feelings in and through her gestures, facial expressions, tone of voice, 'body language' and so forth. We do not initially see the 'bare' behaviour that a physiologist might describe as simply the movements of limbs and muscles, and then infer from that the existence of mental states in a hidden private realm, postulating these as the probable causes of the bodily movements. Rather, we experience the bodily behaviour straight off as expressive of mind; we just see the gesture as one of annoyance of worry or whatever. Here, body and mind are connected expressively in that the former incarnates or makes manifest the latter; and the individual gestures, facial expressions and so on that the person makes are all connected together by virtue of the fact that they together all serve to express the underlying state of mind. For it is characteristically a *pattern* of behaviour that makes manifest an emotion.

It should, I hope, already be obvious from this that Wittgenstein's *later* philosophy stands squarely in the expressivist tradition, as least as far as its understanding of persons and its critique of dualism and materialism is concerned. I want to suggest, however, that this expressivist outlook is already present in the *Tractatus*, and, indeed, in a metaphysically more ambitious way. In his early work, Wittgenstein sees not just individual human beings, but the world as a whole in expressive terms. Anscombe puts the point nicely:

> There is a strong impression made by the end of the *Tractatus*, as if Wittgenstein saw the world looking at him with a face...he speaks of the world 'waxing and waning as a whole' i.e. in terms of my analogy, as having more or less expression, or a good or evil expression.[1]

In seeing the world in this way, Wittgenstein was following Schopenhauer. For both of them, the phenomenal world itself is expressive of an underlying metaphysical reality that they call Will. Let us see what this means in Schopenhauer's work, before returning to Wittgenstein.

Schopenhauer was a Kantian, and therefore held that the empirical world of causally related spatio-temporal substances existed only relative to the perceptual and conceptual apparatus of conscious subjects. '"The world is my representation": this is a truth valid with reference to every living and knowing being.'[2] However, apart from its phenomenal existence for us, the world exists in itself. Schopenhauer departs from Kant's agnosticism about the thing-in-itself, identifying it as force or energy or – as he, perhaps unfortunately, puts it – Will. I know myself in a quite different way from any other object in the phenomenal world – but I do not know myself as pure mind or intellect, but primarily as will, striving, desire. Extrapolating from this self-knowledge, I arrive at the conclusion that the inner essence of all things is similarly a striving or energy, although in most cases unaccompanied by consciousness. He who sees this

> will recognise that same will not only in those phenomena that are quite similar to his own, in men and animals, as their innermost nature, but continued reflection will lead him to recognise the force that shoots and vegetates in the plant, indeed the force by which the crystal is formed...and finally even gravitation, which acts so powerfully in all matter, pulling the stone to the Earth and the Earth to the Sun; all these he will recognise as different only in the phenomenon, but the same according to their inner nature. He will recognise them all as that which is immediately known to him so intimately and better than everything else, and where it appears most distinctly is called will.[3]

Schopenhauer thus maintains a strongly anti-dualist outlook; my body is my will, known from the outside. Thus there is no causal nexus linking my acts of will and my bodily movements; the act of will and the movement are one and the same, but merely known in different ways.[4] More generally, the world as Will and as Representation are not distinct entities; the latter is simply the former as perceived by us. Just as my actions express my feelings, intentions and desires, so whatever happens in the phenomenal world expresses the restless striving of the universal Will. (of which, indeed, my own will is itself merely an expression).

Regarded as an explanatory theory, Schopenhauer's philosophy faces acute difficulties. (Not, of course, that this distinguishes it from any other philosophical system.) How is it, for instance, that the universal Will separates itself out into numerous discrete perceivers, who then constitute the world as representation? Especially considering that multiplicity is, as Schopenhauer stressed, only a feature of the world as representation? How is it that the Will manages, as it were, to develop consciousness which can then look back on the indivisible reality of which they are part? And how can the noumenal, which is outside time, be will or striving? Can we understand willing outside time? It is in any case very doubtful whether, given his Kantian premises, Schopenhauer is really entitled to go beyond Kant's agnosticism about the thing in itself.

Regarded as an expressive vision, however, Schopenhauer's philosophy is enormously impressive, and it is not surprising that many of the greatest European writers and artists of the late nineteenth and early twentieth centuries were fascinated by Schopenhauer. For his metaphysics is not – and was not supposed by him to be – a strictly logical system, in which conclusions follow from indisputable premises with rigorous necessity. Rather, it is based on an analogy between our own inner being and that of the world as a whole that we may come to experience as compelling. If one comes to accept Schopenhauer's philosophy, it is not as the result of following a purely logical chain of deductions; it is through leaning a new way of seeing the world. The familiar objects of perception are no longer experienced as dead mechanisms, but as expressions of the same restless striving that I feel within myself.

> The force that through the green fuse drives the flower
> Drives my green age; that blasts the roots of trees
> Is my destroyer.
> And I am dumb to tell the crooked rose
> My youth is bent by the same wintry fever.[5]

II

There can be no doubt that Wittgenstein was strongly influenced by Schopenhauer; the 1916 *Notebook*, where many of the later propositions of the *Tractatus* were first formulated, reads in considerable measure as a dialogue with Schopenhauer. Pears, however, insists that, despite this, Wittgenstein 'never accepted Schopenhauer's vision of the world as a manifestation of will'.[6] I think Pears is wrong about this, and consequently wrong in his interpretation of the *Tractatus* as essentially realistic. In a 1916 *Notebook* entry, Wittgenstein writes: 'I can also speak of a will that is common to the whole world. But this will is in a higher sense my will. As my representation is the world, in the same way my will is the world-will'(NB 85).[7] This is why he can say 'The world and life are one' (TLP 5.621) and 'I am my world (the microcosm)' (TLP 5.63). Wittgenstein accepts Schopenhauer's distinction between the world as Will and the world as Representation, but he differs in the account he gives of the latter. For Wittgenstein, representation is essentially a linguistic matter; the world, as it is for us, is the world as we represent it to ourselves in propositions. Hence the main task of the *Tractatus* is to explain the general form of the proposition – to set out the basic logical scaffolding upon which any language must be constructed. And this will determine how I will see the world insofar as I think about it in language. Hence 'the limits of language (of that language which alone I understand) mean the limits of my world' (TLP 5.62). To apprehend the world by means of propositions that are truth-functions of elementary propositions, which are themselves structured assemblages of simple signs, is to see the world as a concatenation of contingent, mutually independent states of affairs, these being combinations of simple objects, which are 'unalterable and subsistent' (TLP 2.0271).

'To give the essence of a proposition means to give the essence of all description, and thus the essence of the world' (TLP 5.4711). Here we can clearly see the welding together of two seemingly very different traditions of thought, which is one of the most fascinating features of the *Tractatus*. That the essence of the proposition gives us the essence of all description is the Fregean 'linguistic turn', the move from a psychological analysis of thought to a logical analysis of language. The second move, from the essence of description to the essence of the world, is the idealist – specifically Schopenhauerean – move. For the essence of the world only follows from the essence of description on the idealist assumption that the world has no reality apart from our description of it. The ontology of the *Tractatus* has always puzzled the commentators; it is deduced from the structure of language, but, if one adopts a realist interpretation, it is simply a matter of metaphysical luck that the world has this structure which enables it to correspond to what Wittgenstein thought of as the necessary structure of any language. This seems embarrassingly implausible, and so some commentators (for example, McGuinness) have taken the ontology as merely an expository myth, a way of presenting the argument about the structure of language: 'An object in the *Tractatus*, which is the reference of a name or simple sign, can be viewed as simply the truth-value potential of a certain expression.'[8]

I think that this is more on the right lines, but I would not describe the *Tractatus* objects as mythical or as rhetorical fictions. Wittgenstein is concerned with the nature of the world, not just the nature of language, but it is the world as it is for us which he attempts to describe. Applying the linguistic turn to the Kant/Schopenhauer tradition, Wittgenstein gives us neither Rationalist ontology (deducing the nature of Reality in itself from the structure of language) nor simply a description of the structures of language; he gives us an ontology, but one which is relative to language. Since he thinks that there is a single specifiable essence of language, this is an ontology which must be adopted by all language-users; there is no conceivable rival with which it might be compared. So, like Kant, he avoids relativism and historicism by postulating a basic timeless universal structure of representation, which determines the way in which we must interpret reality to ourselves. The simple objects of Wittgenstein's ontology then do exist, but not as things-in-themselves. Rather, they are presupposed as comprising the basic structure of the world as (linguistic) representation.

III

The world, as it is for us, then, is a collection of contingent facts. None of them has in itself any necessity (TLP 6.37) or value (TLP 6.41). All that literally meaningful language can do is attempt to picture these facts. But Wittgenstein is an expressivist, not a positivist; for him, what cannot be stated or apprehended as factual may nevertheless 'make itself manifest' in and through the facts. The doctrine of 'showing', as opposed to saying, originated as a logical one. The very first sentence in the *Notebooks* is 'Logic must take care of itself' (NB 2). This is (almost) repeated at TLP 5.473, and at TLP 4.0312 Wittgenstein tells us that his 'fundamental idea' is 'that the "logical constants" are not representatives'. Logic is not a description of any reality – it must take care of itself because there is no independent reality corresponding to it which could take care of it. Logic is radically different from factual discourse. If logic were – as Frege and Russell had both thought – descriptive of a realm of facts, no matter how abstract, it could not have the necessity it does. Any reality – any set of facts – is contingent; it could be other than it is. But logic cannot be about any set of facts, because it is prior to all facts. It is the condition of factuality: 'A statement cannot be concerned with the logical structure of the world, for in order for a statement to be possible at all, in order for a proposition to be CAPABLE of making SENSE, the world must already have the logical structure that it has. The logic of the world is prior to all truth and falsehood' (NB 14).

Logical propositions do not describe logical facts; they are merely tautologies (TLP 6.1). Logic, as something set out in textbooks, is part of the ladder that we must kick away in order to climb up beyond it to enlightenment (TLP 6.54). 'Logic is not a body of doctrine', it is 'transcendental' (TLP 6.13).

The Kantian word here is interesting. In the *Notebooks* Wittgenstein calls both ethics and logic 'conditions of the world' (NB 77). And it is primarily these that can be shown, but not said. As James Edwards has put it, 'showing is an escape hatch from the realm of nonsense'.[9] The things that can be shown but not said are those whose expression in literally meaningful language is ruled out by the *Tractatus*' highly restrictive theory of meaning, but which cannot be discarded as simply absurd.[10] We can only speak of contingencies, but neither ethics nor logic can be treated as contingent. They are therefore to be regarded not as facts within the world, but as limits or conditions of it. And those conditions are not wholly ineffable; they can be indicated.

Logical form cannot itself be spoken of directly, but it expresses itself in the ordinary propositions of our language. Any banal everyday remark anyone makes already makes manifest what logical form is, for 'all the propositions of our everyday language, just as they stand, are in perfect logical order' (TLP 5.5563). One cannot talk about logic, since any talking already presupposes logic:

> Propositions cannot represent logical form: it is mirrored in them.
> What finds its expression in language, language cannot represent.
> What expresses ITSELF in language, WE cannot represent by means of language.
> Propositions SHOW the logical form of reality.
> They display it. (TLP 4.121)

Wittgenstein refers here to the logical form 'of reality', not just of language. This is, as I have argued above, reality as apprehended in language, the world as (linguistic) representation. Because description is linguistic, and logic is the essence of language, it follows that logic is the transcendental condition for the World as Representation and, precisely for that reason, it is not something that can itself be represented. It must be left to make itself manifest in the logical structure of the world as we apprehend it linguistically.

IV

However, the notion of showing does not only apply to the logical form of propositions showing itself in those propositions. It also applies to the sense of life showing itself in the facts of the world. For Wittgenstein, as for Schopenhauer, the world as it really is, the world as Will, makes itself manifest in the world as Representation. The facts are in themselves without value, but if they are seen together, *sub specie aeterni* (TLP 6.45), they can be seen as expressing the underlying noumenal Will. In the *Notebooks*, Wittgenstein – here departing radically from Schopenhauer – identifies the Will as divine. 'How things stand is God. God is, how things stand' (NB 79). However, Wittgenstein does not simply identify God with the totality of facts; rather, He is what they express. When one sees the facts together in the right way, one sees them as the manifestation of the divine will.

In the *Tractatus* itself, he seems to distinguish sharply between God and the world; '*How* things are in the world is a matter of complete indifference for what is higher. God does not reveal himself *in* the world' (TLP 6.432). What I think this means is that God cannot be thought of as one being amongst others (even if the 'Supreme Being'); He cannot be seen as revealed more in any one fact or set of facts than another. It is the world as a whole, rather than any set of facts in it, which manifests God.

> The facts all contribute only to setting the problem, not to its solution.
> It is not *how* things are in the world that is mystical, but *that* it exists.
> To view the world sub specie aeterni is to view it as a whole – a limited whole.
> Feeling the world as a limited whole – it is this which is mystical. (TLP 6.4321–6.45)

To feel the world as a limited whole is to be aware of another dimension of reality by which the phenomenal world is limited. 'The solution of the riddle of life in space and time lies *outside* space and time' (TLP 6.4312). For Schopenhauer and Kant, the noumenal lies outside space and time. If I am right that Wittgenstein identifies God with Schopenhauer's noumenal Will, then God is indeed beyond time and space, and beyond the limits of a language which can only denote contingent facts. But this whole world which we experience as a concatenation of contingencies is, in its deepest essence, identical with the divine will. God is not, therefore, apart from the world; He *is* the world. Of course, according to the *Tractatus* itself, we can't really say that. Nevertheless, it can make itself manifest to us; that is, we can come to experience the world 'mystically', as an expression of God's will. This is not a matter of seeing any particular fact(s) differently, but of seeing them all together and recognizing their pattern as expressive of something that could not be seen in any one individually. As one may step back from an apparently random array of dots, and see them as the picture of a face ('...as if Wittgenstein saw the world looking at him with a face'). Wittgenstein's outlook at the time of the *Tractatus* could be called 'pantheistic'; the facts of the world are not just seen as brute facts, but as expressive of the divine will and, as such, to be accepted with equanimity whatever they may be.

If the world is the manifestation of a divine, noumenal will, what of the human will? Wittgenstein distinguishes between the will 'as a phenomenon...of interest to psychology' (TLP 6.423) and the ethical will, which is not of interest to any science, since it cannot really be spoken of. What does this distinction amount to? In the *Notebooks* Wittgenstein writes: 'The thinking subject is surely mere illusion. But the willing subject exists. If the will did not exist, neither would there be that centre of the world, which we call the I and which is the bearer of ethics' (NB 80).

The rejected thinking subject is the Cartesian soul. There is no such unitary thing; just a set of discrete, though causally related, thoughts. But Wittgenstein's premises no more allow for a willing than for a thinking subject in the world. So the willing subject that is affirmed can only be a limit of the world. For Wittgenstein the will is essentially connected to ethics. 'I will call "will" first and foremost the bearer of good and evil' (NB 76). A few days later, he adds, 'Good and evil only enter through the subject. And the subject is not a part of the world, but a boundary of the world' (NB 79). So we have the notion of the willing subject, the bearer of good and evil, as the limit of the world.

What does this mean? Wittgenstein rejects the Cartesian self, but doesn't draw the Humean conclusion that all there is is a bundle of discrete mental facts. Admittedly, that is all there is *in* the (phenomenal) world, but there is also the condition of there being a phenomenal world at all – the Transcendental Ego. This is not a fact in the world, but the 'fact' (or meta-fact?) that the world is experienced – and therefore experienced as a concatenation of contingent atomic facts – at all. This is, of course, yet another of the things that cannot literally be said at all. When we attempt to articulate it, we fall into the incoherent doctrine of solipsism. But solipsism is significant nonsense: 'What the solipsist *means* is quite correct; only it cannot be *said* but makes itself manifest' (TLP 5.62). As phenomena ourselves, we are just small and insignificant parts of the world. But we are not *just* phenomena; it is only in us that the phenomenal world is constituted at all. We are not just objects in the world, for each of us plays the role of the Transcendental Ego or, as Schopenhauer puts it, the Subject, which is

> that which knows all things and is known by none...the supporter of the world, the universal condition of all that appears, of all objects, and is always presupposed....Everyone finds himself as this subject, yet only insofar as he knows, not insofar as he is object of knowledge.[11]

As the last sentence makes clear, 'the Subject' for Schopenhauer denotes not an individual – a mysterious world-constituting entity – but a function. The subject is each of us, considered just barely as a knowing and sensing mind. It only takes one such mind, equipped with the Kantian array of *a priori* forms of intuition and categories, to constitute the world as Representation, for that simply *is* the noumenal world, as it is known by such a mind.

Wittgenstein is thus following this Kantian/Schopehauerean line when he writes: 'The subject does not belong to the world, rather, it is a limit of the world' (TLP 5.632). It is, however, rather peculiar that Wittgenstein rejected the thinking subject while retaining the willing subject. There is neither a thinking nor a willing subject *in* the world, but there is the metaphysical subject, as the limit of the world. And this subject, it seems, both thinks and wills. In the former role it constitutes the world as Representation, in the later it adopts an attitude towards it. Hence the pronouncement: 'Ethics must be a condition of the world, like logic' (NB 77). Ethics is the condition of the world viewed by the willing subject, logic that of the world viewed by the thinking subject. The exercise of the will in this sense is not a fact in the world, but an attitude that is taken towards the world as a whole. And it is this that can be the subject of ethics. Acts of will, like thoughts, actions or any other facts, are all on the same level; they are without value, they just are (TLP 6.41). If we are to have a place for ethics, it cannot be within the pattern of facts that comprises the world, but only in the fundamental attitude that I adopt towards the world.

> If the good or bad exercise of the will does alter the world, it can alter only the limits of the world, not the facts – not what can be expressed by means of language.
> In short, the effect must be that it becomes an altogether different world. It must, so to speak, wax and wane as a whole.
> The world of the happy man is a different one from that of the unhappy man. (TLP 6.43)

In talking of happiness and unhappiness here, Wittgenstein is not just giving an example; the words have a particular meaning for him. 'In order to live happily, I must be in agreement with the world. And that is what being happy MEANS' (NB 75). The happy man is one who accepts whatever comes his way as the will of God, who is thus able to be content, come what may. The unhappy man is one whose striving bring him into conflict with the way things are; he cannot accept the world however it may be. Rather, as soon as things start to go wrong for him – and nothing he can ever do can guarantee that they will not – he experiences the world as oppressive. The facts seem to be hemming him in, stifling him; the 'face' with which the world looks out at him has become menacing, ugly.

I shall not attempt to provide an estimate of this Stoical quietism that Wittgenstein recommends as an ethical outlook, although it seems to have some obvious and worrying flaws. Although Wittgenstein accepts the noumenal Will as divine, while Schopenhauer considers it the source of endless suffering and misery, the two philosophers both recommend a rather similar ethical stance. For both, to will particular things is to expose oneself to the misery of having one's desires frustrated, as well as bringing one into conflict with others. Wittgenstein recommends abandoning one's personal will, by accepting whatever fate brings; Schopenhauer also commends the abandonment of willing, but notes that for a being which is essentially will, this is tantamount to self-annihilation.

V

I have been talking about an expressive vision of things. But is this the normal way we experience the world, at least when we refrain from blinding ourselves with scientistic or metaphysical philosophies, or is it a kind of rare, visionary experience? For Wittgenstein, to be happy or unhappy are both expressive responses to the world; the facts, when taken together as a whole, are seen not just as neutral and value-free, but as providing a significant context for my life. Schopenhauer would presumably be an example of an unhappy man whose will is not in agreement with the world; he sees it as significant, as expressive – but expressive of a horror and destructiveness from which he recoils. But it also seems that there can be an experience of the world that is expressively dead – one in which it is simply seen as a collection of brute facts. 'The world of the happy is a happy world. Then can there be a world that is neither happy nor unhappy?' (NB 78). The growth of scientific rationalism and industrialization have brought about what Weber called the 'disenchantment' of the world, the tendency to simply see it as a dead mechanism, a collection of value-free facts, as a stockpile of resources to be exploited. It was, of course, against this process and its intellectual reflection in the philosophy of the Enlightenment that the Romantic expressivists were reacting. Wittgenstein was not a Romantic, but he frequently expressed his sense of distaste for, and unease in, contemporary civilization. He had a strong primitivist, Tolstoyan streak, which led him to work as a village schoolteacher, and, later, to seriously consider going off to work, perhaps as a doctor or even as a manual labourer, in some remote part of Russia.[12] In this, he belongs with a powerful current in post-Romantic Western sensibility which has sought to escape from the false sophistications of a technocratic society and recover a fresher, more primordial vision of things (cf. the widespread fascination with primitive art at the beginning of the twentieth century, and its influence on artists such as Picasso and Matisse, Stravinsky's *Rite of Spring*, D.H. Lawrence's 'savage pilgrimage' to non-European cultures, Surrealism – the list could be considerably extended).

It is indeed to art that we would normally think of turning in order to give an expressive enrichment to our experience. In the *Tractatus* Wittgenstein states that 'Ethics and aesthetics are one and the same' (TLP 6.421). Given that ethics for Wittgenstein at this time is simply a matter of the attitude one adopts to the world, this suggests that aesthetics is also concerned with experiencing things in the right spirit. He talks of seeing things *sub specie aeternitatis*: 'In such a way that they have the whole world as background' (NB 83). Aesthetic perception sees even trivial and ugly things in a way that invests them with significance. I think one can say that aesthetic contemplation considers things in their own right, not as members of a class, not as useful for this or that purpose, but simply as being, and as being what they are.

> If I have been contemplating the stove, and then am told: but now all you know is the stove, my result does indeed seem trivial. For this represents the matter as if I had studied the stove as one among the many things in the world. But if I was contemplating the stove, IT was my world, and everything else colourless by contrast with it. (NB 83)

One may contemplate a stove aesthetically, but one may also turn to art. A successful work of art, one might say, forces us to perceive it aesthetically. As Tilghman, to whose discussion of Wittgenstein's aesthetics I am indebted, puts it, 'the work of art selects an object, a scene, a situation and makes that object stand still to be contemplated and in so doing treats the object as if it were a world unto itself so that it becomes *my world* and a *representative of the whole*'.[13]

This attitude, at once ethical, 'mystical' and aesthetic, is, I think, central to the aim of the *Tractatus*, and carries through, in altered ways to the later work. Wittgenstein's attack on explanation is aimed at recovering a freshness of response to the world that the scientific attitude seems to block. I believe that this – the opening up of the possibility of expressive vision, which should be our normal response to the world but which is occluded in our culture – is what Wittgenstein has in mind when he writes of transcending the propositions of the *Tractatus* in order to 'see the world aright' (TLP 6.54).

Notes

1 G.E.M. Anscombe, *An Introduction to Wittgenstein's Tractatus* (London: Hutchinson, 1959), p. 172.

2 Arthur Schopenhauer, *The World as Will and Representation*, Vol. 1, trans. E.F.J. Payne (New York: Dover, 1969), p. 3.
3 Ibid., p. 110.
4 These ideas perhaps had some influence on Wittgenstein's later philosophy. In the *Notebooks 1914–16* Wittgenstein did develop an account of agency along these Schopenhauerean lines, but concludes by saying that it 'makes it look as if some part of the world were closer to me than another (which would be intolerable)': Ludwig Wittgenstein, *Notebooks 1914–16*, ed. G.H. von Wright and G.E.M. Anscombe, trans. G.E.M. Anscombe, 2nd edn (Oxford: Blackwell, 1979), p. 88. (Subsequent references to this work will be indicated by 'NB' in the text.) Accordingly, in the *Tractatus* itself he gives a much more Cartesian-sounding account of the relation between will and act as a contingent causal one. See Ludwig Wittgenstein, *Tractatus Logico-Philosophicus*, trans. B.F. McGuinness and D.F. Pears (London: Routledge, 1961), 6.373–6.375, hereafter referred to as 'TLP' in the text.
5 Dylan Thomas, 'The Force That Through the Green Fuse Drives the Flower', in *Poems*, ed. D. Jones (London: J.M. Dent, 1982), p. 77.
6 D. Pears, *The False Prison: A Study of the Development of Wittgenstein's Philosophy*, Vol. 1 (Oxford: Clarendon Press, 1987), p. 5.
7 I have altered Anscombe's translation of *Vorstellung* in this passage from 'idea' to 'representation', in order to conform with Payne's usage in his Schopenhauer translations.
8 B. McGuinness, 'The So-called Realism of the *Tractatus*', in I. Block (ed.), *Perspectives on the Philosophy of Wittgenstein* (Oxford: Blackwell, 1981), p. 65.
9 J. Edwards, *Ethics Without Philosophy: Wittgenstein and the Moral Life* (Tampa: University Presses of Florida, 1982), p. 15.
10 The standard view that Wittgenstein is operating with a category of 'significant nonsense' in the *Tractatus* has been challenged by Cora Diamond, who claims that Wittgenstein has no such category and always simply uses 'nonsense' to mean simply gibberish. See her *The Realistic Spirit* (Cambridge, MA: MIT Press, 1991) especially Chapters 3, 6 and 8. I don't have space to consider her account here; for some effective criticism, see J. Lippitt and D. Hutto, 'Making Sense of Nonsense: Kierkegaard and Wittgenstein', in *Proceedings of the Aristotelian Society*, **XCVIII** (3), 1998, pp. 263–86. I shall take it in what follows that 'nonsense' does not always just mean gibberish for Wittgenstein; in particular, he clearly does not want to dismiss ethics and religion as simply absurd, but at the time of the *Tractatus* he did describe them as 'nonsense'. These facts can hardly be reconciled unless we do admit some category of 'significant nonsense', and then there can be no objection in principle to extending the category further.
11 Schopenhauer, *The World as Will and Representation*, *op. cit.*, n. 2, p. 5.
12 Ray Monk, *Ludwig Wittgenstein: The Duty of Genius* (New York: Free Press, 1990), pp. 347–54.
13 B.R. Tilghman, *Wittgenstein, Ethics and Aesthetics: The View From Eternity* (Basingstoke: Macmillan, 1991), p. 54.

- Pretty trad. Kenniston-type of a reading. (No questions about nonsense, or explaining it.)
- W's an ~~[illegible]~~ ling. idealist (in contr. to a psychologist one)
- Takes into acc. also ethics + aesthetics.

Chapter 3

To What Extent is Solipsism a Truth?

Michael Kremer

My title[1] is taken from one of the most obscure, and most discussed, sections of an already obscure and much discussed work, the discussion of the self, the world, and solipsism in sections 5.6-5.641 of Ludwig Wittgenstein's *Tractatus Logico-Philosophicus*.[2] Wittgenstein writes:

> 5.6 *The limits of my language* mean the limits of my world.
>
> 5.61 Logic fills the world: the limits of the world are also its limits.
> We cannot therefore say in logic: This and this there is in the world, that there is not.
> For that would apparently presuppose that we exclude certain possibilities, and this cannot be the case since otherwise logic must get outside the limits of the world: that is, if it could consider these limits from the other side also.
> What we cannot think, that we cannot think: we cannot therefore *say* what we cannot think.
>
> 5.62 This remark provides a key to the question, to what extent solipsism is a truth.
> In fact what solipsism *means*, is quite correct, only it cannot be *said*, but it shows itself.
> That the world is *my* world, shows itself in the fact that the limits of the language (*the* language which I understand) mean the limits of *my* world.

Interpretations of the ensuing passages have tended to focus on solipsism as an epistemological or metaphysical thesis.[3] While such discussions have their value, and I will draw on them at points in my own reading of the text, I want to focus on another way of approaching these passages. This approach is suggested by a remark of Brian McGuinness, that 'solipsism ... in his [Wittgenstein's] case was not an intellectual exercise but a moral and mystical attitude'.[4] I think the general thrust of McGuinness' remark is correct, although I would not put the point quite as he does. I would prefer to say that solipsism is an intellectual, moral and mystical exercise aimed at bringing about a change in one's spiritual life. Simone Weil once wrote 'subjectivism, absolute idealism, solipsism, scepticism ... the Upanishads, the Taoists and Plato ... all of them, adopt this philosophical attitude by way of purification'.[5] I would add Wittgenstein to this list; and so I want to construct a reading of *Tractatus* 5.6-5.641 which will help to illuminate the ethical purpose which Wittgenstein famously claimed for the book.[6]

I will argue, however, that to *stop* with the purification achieved through solipsism leads to an incomplete and misleading conception of Wittgenstein's ethical vision. James C. Edwards, in his book *Ethics Without Philosophy*, and Emyr Thomas, in a recent article, 'From Detachment to Immersion: Wittgenstein on the Problem of Life', develop interpretations of the *Tractatus*' ethical outlook which are, to my mind, incomplete in this respect.[7] Both Thomas and Edwards end up accusing Wittgenstein, in the *Tractatus*, of a failed attempt at selflessness, an essentially narcissistic, heroic vision, of the self as perfecting itself through detachment.[8] In contrast, I will argue that the ultimate point of the *Tractatus*' discussion of solipsism is that detachment is incomplete unless, and until, it ends in a complete and final self-abandonment, in which the very conception of the 'I' required by solipsism is left behind. This rejection of all forms of self-assertion, including the self-assertion found in certain misguided forms of asceticism, piety and false humility, is what the *Tractatus* really aims at. In this respect, the *Tractatus* stands in the great tradition of mystical thought, and in this paper I will try to say something about the way in which Wittgenstein can be placed in that tradition. After a meeting with Wittgenstein in December 1919, Russell wrote to Lady Ottoline Morell: 'I had felt in his book a flavour of mysticism, but was astonished when I found that he has become a complete mystic. He reads people like Kierkegaard and Angelus Silesius. ... It all started from William James' *Varieties of Religious Experience* ... He has penetrated deeply into mystical ways of thought and feeling ... .'[9] My analysis of the *Tractatus*' discussion of solipsism will help to make clear the truth in these remarks.

I

My project, then, is to develop a reading of the *Tractatus*' remarks on solipsism from the perspective of the ethical aim of the work. In order to give some initial plausibility to this project, I want to say something about the terms employed in the key passage, TLP 5.62:

> This remark provides a key to the question, to what extent solipsism is a truth.
> In fact what solipsism *means*, is quite correct, only it cannot be *said*, but it shows itself. (TLP 5.62)

The three terms are: 'truth', 'means' and 'shows itself'. I want to give each of these terms, as they occur in TLP 5.62, a *practical* reading.

I begin with 'shows itself'; once this is properly in place the other two terms will be easily dealt with. The *Tractatus* famously draws a distinction between that which can be *said* in language and that which can only be *shown*. For example, Wittgenstein argues that the 'logical form' which a proposition shares with the reality that it depicts cannot itself be made the subject of depiction, but is *shown* in the proposition which possesses it. However, talk of logical form as 'shown' in a proposition easily tempts us into a picture in which 'logical form' is a *something*, an entity standing in relation to a proposition, so that it a *fact* that this proposition has that logical form. If our language cannot depict this fact, then it seems this must be a merely contingent limitation which can be overcome by rising to the level of a meta-language, as Russell suggested in the Introduction to the *Tractatus*. The Tractarian response to such a suggestion will be that in discussing 'logical form' we are attempting to state the essential presuppositions of any meaningful language whatsoever. To treat this as a *fact* which could be represented in a language is to suppose a language in which we could at least envisage the possibility of this fact's *not* obtaining. But such a language, Wittgenstein would argue, would be an 'illogical language' – which is to say an impossibility.

Yet the *Tractatus* itself, insofar as it helps itself to talk of 'showing' logical form, seems to be such an impossible language. Notoriously, Wittgenstein himself not only accepts this charge, but explicitly applies it to his own work:

> My propositions are elucidatory in this way: he who understands me finally recognizes them as nonsense, when he has climbed out through them, on them, over them. (He must so to speak throw away the ladder, after he has climbed up on it.)
> He must surmount these propositions; then he sees the world rightly. (TLP 6.54)

The proper understanding of this passage has been the subject of much debate. One prominent line of interpretation, represented by such commentators as P.M.S. Hacker and David Pears, has it that Wittgenstein recognizes a realm of ineffable fact-like quasi-truths, such as the fact-like quasi-truth that language and world share a common logical form. These quasi-truths cannot be expressed in language, but are *shown* in the proper workings of language. On this view, the propositions of the *Tractatus*, while literally nonsense, serve a useful purpose by directing our attention to the ineffable features of reality and language that undergird all our meaningful discourse. In this way, they enable us to 'see the world rightly'.

This line of thought has been attacked by such interpreters of the *Tractatus* as Cora Diamond and James Conant.[10] Diamond calls it 'chickening out', a refusal to take seriously Wittgenstein's claim that his propositions are nonsense. She insists on a 'resolute' reading of this claim. The propositions of the *Tractatus*, insofar as they attempt to circumscribe the limits of language and thought, are 'simply nonsense', as the Preface tells us:

> The book will... draw a limit to thinking, or rather – not to thinking, but to the expression of thoughts; for, in order to draw a limit to thinking we should have to be able to think both sides of this limit (we should therefore have to be able to think what cannot be thought).
>
> The limit can, therefore, only be drawn in language and what lies on the other side of the limit will be simply nonsense.

As Diamond argues, to take the propositions of the *Tractatus* to gesture at ineffable quasi-facts underlying all factual language is to continue to image necessity as a fact; it inevitably leads to the nonsense of formulating as propositions that which is said to be inexpressible. Hacker's discussion of solipsism provides some prime examples of this difficulty. He is driven into saying things like this:

> What the solipsist means, and is correct in thinking, is that the world and life are one, that man is the microcosm, that I am my world. These equations... express a doctrine which I shall call Transcendental Solipsism. They involve a belief in the transcendental ideality of time. ... Wittgenstein thought that his transcendental idealist doctrines, though profoundly important, are literally inexpressible.[11]

The problem here is of course, that it is only by expressing it that Hacker manages to seem to communicate to us what it is that is supposed to be inexpressible. In attributing such ineffable doctrines to Wittgenstein, Hacker ignores Wittgenstein's characterization of philosophy as 'not a theory but an activity' whose result is 'not a number of "philosophical propositions", but to make propositions clear' (TLP 4.112). He takes 'we cannot ... say what we cannot think' (TLP 5.62) to express a thesis of the 'inexpressibility of the unthinkable'.[12] But he ignores the *Tractatus*' claim that 'the thought is the significant proposition' (TLP 4) with its strong suggestion that the inexpressible is, conversely, unthinkable.

Now I agree with Diamond and Conant that irresolute readings such as Hacker's are, at bottom, incoherent nonsense; and it is a methodical purpose of the *Tractatus* to bring us to see that philosophical theorizing, conceived as offering us a source of grounding or justification for logic, language or life, will, in the end, produce nothing but such nonsense. Nonetheless, I do not think that all deployments of the terminology of 'showing' in the *Tractatus* are *irredeemable* nonsense. There is a sense that can be given to some of these uses of 'showing' which does not degenerate into the incoherence of envisaging in the form of a fact that which we declare not to be a fact. We should *not* read talk of 'showing', and correlatively of 'perceiving', 'seeing' and 'recognizing' that which is shown on the model of a *relation* between a subject and some ineffable fact-like entity ('that *p* is shown to *S*', '*S* perceives that *p*'). This form of the idea of showing is exactly what the *Tractatus* wants to teach us to abandon. Rather, we should read talk of 'showing', and correlatively 'seeing', on the model of the demonstration of a technique and the uptake required to understand the demonstration. In essence, my suggestion is that one who 'sees' that which is shown, is simply one who 'knows how to go on'.[13] On this reading, phrases such as 'the logical form of the world' have no independent meaning; it is only the larger contexts – 'showing the logical form of the world', 'seeing the logical form of the world' and so on – which have a meaningful use. These phrases are like 'dancing the waltz', which should be understood simply as 'waltzing' rather than as involving a relation between a dancer and a particular 'dance' (the waltz). It is significant that the introduction of the terminology of 'showing' at TLP 4.022 ('A proposition shows its sense') is embedded in a discussion of *understanding*. To understand a proposition is to 'know what is the case, if it is true' (TLP 4.024). However this should not be seen as an instance of 'knowledge-that'. To understand a proposition *p* is not to know another proposition of the form '*p* is true if and only if *q*'. Clearly knowing such a proposition presupposes understanding *p* and cannot explain it. Rather, understanding is a form of 'knowledge-how'. One understands a proposition by knowing how to use it – when to assert it and when to deny it.[14]

In general, then, I would suggest that uses of 'showing' in the *Tractatus* may be two-sided.[15] On the one hand, talk of showing can tempt us into the nonsensical illusion that we grasp a realm of super-facts beyond the reach of language. On the other hand, talk of showing can, innocently enough, direct us to the practical abilities and masteries that are part of our ongoing talking, thinking and living.

If we can redeem, in some contexts, the notion of 'showing' in this fashion, then I think it is also possible to redeem talk in the *Tractatus* of the communication of a 'truth' which is other than propositional. But, again, we should not be drawn into the nonsensical illusion of a realm of ineffable proposition-like 'truths' which we can yet, mysteriously, grasp. There is another sense of 'truth' which we can appeal to here, exhibited in such Biblical passages as: '...whoever does the truth comes into the light... .'[16] 'Truth' as something which one can *do* is not something we might be tempted to think of as expressible in a proposition. It is, rather, a way to be followed, a 'path' for life. Insofar as the *Tractatus* communicates a 'truth' it is by demonstrating to us, in practice, how to follow such a path.[17] Solipsism, I want to argue, is *part* of this path; and in asking 'to what extent solipsism is a truth?' Wittgenstein is asking *how far* solipsism will take us along the way. This aspect of Wittgenstein's question – '*How far* is solipsism a truth? – is typically submerged in more metaphysically and epistemologically oriented readings of the remarks on solipsism, and it is one of the advantages of my way of understanding these remarks that it makes room for a dimension along which we can, in some sense, 'measure' the truth of solipsism.

Now Wittgenstein's general answer to our question is this: 'In fact what solipsism *means* is quite correct, only it cannot be *said*, but it shows itself.' We have already seen how I intend to read the claim that 'what solipsism *means*' *shows itself* and is *correct* – that is, a *truth*. But in what sense can this be what solipsism 'means'? Does not talk of 'meaning' here suggest, after all, that solipsism is a (failed) attempt to put across a true doctrine – something like a linguistic proposition? My response to this corresponds to my reading of 'showing' and 'truth'. The word translated as 'means' by Ogden ('*meint*'), has various senses, as does the English 'means'. One of these is 'intends' as in '*Sie meint es gut*' ('She means well'). It is in this sense of 'what solipsism intends' that I would like to read 'what solipsism means', not in the sense of 'what solipsism intends to *say*', but rather in the sense of the underlying intention of solipsism. It is this intention that is said to be 'quite correct' and, so far as it goes, a truth.

Putting all this together, I suggest the following reading of 5.62:

> This remark provides the key to the question as to how far solipsism can take us along the path of spiritual and ethical enlightenment. In fact, the intention of solipsism is a good one, but this cannot be communicated through a set of principles, but must be demonstrated in practice.

My goal is to substantiate this reading by showing how it is possible to read the ensuing discussion of solipsism as a set of instructions for a kind of spiritual exercise or, better, to show how the writing and reading of this discussion, and of the book that contains it, can constitute such an exercise. As a preliminary to this I need to say something about one more concept deployed in TLP 5.62, the concept of the 'limits' of language and the world. Both the numbering scheme of the *Tractatus* and the continuation of TLP 5.62 indicate that the 'remark' which provides the key to the question of the truth of solipsism is TLP 5.6:[18] '*The limits of my language* mean the limits of my world.'

Interpreters have taken widely different views of the role of the concept of 'limits' in the *Tractatus*. In my opinion, the way in which this question is answered will be decisive not only for one's understanding of the metaphysical and epistemological ambitions of the *Tractatus*, but also for one's understanding of its ethical vision. Those interpreters who end up saddling the *Tractatus* with a narcissistic ethics of heroic self-perfection arrive at this point because they have failed to appreciate that the book in fact embodies a thorough-going deconstruction of the notion of 'limits' of language, thought, and world – and so also of the 'self of solipsism'.

II

As we saw above, Hacker's irresolute reading of the *Tractatus*' talk of saying, showing, and nonsense, forces him to *say* that which he takes Wittgenstein to categorize as inexpressible, thereby attributing to Wittgenstein a view which is self-refuting in a fairly obvious fashion. In the context of the discussion of solipsism in particular, one source of Hacker's difficulties is the fact that he takes seriously the notion of 'limits' of language, logic and the world as deployed in TLP 5.6-5.641. Hacker thinks that it makes sense to ask what the limits of language and the world *are* and what the source of delimitation *is*; his answer is that *the* limit, the source of all limits, is the metaphysical subject. This subject is identified by Hacker with a Schopenhauerian metaphysical will, the bearer of good and evil, and also the determiner of the meaningfulness of language. On this view, it is the metaphysical will which establishes the projective relation between language and the world whcih is necessary for names to mean objects and so for propositions to depict states-of-affairs. This produces a tightly delimited Russellian language of experience: 'the sketchy account of "projection" of logico-syntactic forms which Wittgenstein propounded in the *Tractatus* is strongly egocentric on the one hand, and concerned with a "momentary" language on the other.'[19] It is such a view of the metaphysical subject as the determiner of both linguistic meaning and ethical value which leads to the conception of Wittgenstein's ethical outlook as a narcissistic heroism of self-perfection, as we will see.

It is characteristic of resolute readers on the other hand, to see the *Tractatus* as aiming not at determining the source and extent of the limits of language at all, but rather at leading us to see the search for such limits as itself illusory. As Juliet Floyd puts it:

> the book repeatedly conjures up the idea of language and thought as limited, as bounded by a fixed and necessary framework of logical or propositional form, and it also conjures up the idea of logic as limited, as bounded by its role as an inexpressible yet necessary form of infinite complexity mirroring the necessary structure of the world. But the underlying intent of such conjuring ... is to shake us free from these ideas of effability and necessity.[20]

Cora Diamond portrays Wittgenstein's claim that 'the limits of my language mean the limits of my world' as expressing a rejection of a Russellian 'two limits view'. For Russell, the limits of my experience, and so the limits of the objects which I can directly *name*, are narrower than the limits of the world, of all the objects that there are. It is only because I can use descriptions – that is, quantifiers – that I can go beyond the limits of my experience, and reach out towards the limits of the world. The *Tractatus*' conception of meaningful propositions as embedded in a single logical space, and the associated account of the logical functioning of quantifiers, Diamond argues, entail the incoherence of this view. She concludes:

> Wittgenstein's remarks about the limits of language and the world... are concerned with the difference between a Russellian two-limits view ... and a one-limit view. ... The world is *my* world in the sense that there is nothing... which is in *the* world and which *I* cannot name. The idea that the use of quantifiers enables me to reach beyond the limits of my experience to objects 'outside' experience is incoherent. The rejection of the two-limit view does not, though, leave us with one of a sort of thing, namely a limit, of which the Russell view had had two. It is the mistake of solipsism to treat its rejection of a two-limit view as leaving us confined within the limit which Russellian realism had sought to get us beyond. That is, solipsism rejects the Russellian idea that we can get beyond the 'limits of private experience' but keeps its conception of that limit: it does precisely give us one of what Russell had given us two of. The solipsist does not rigorously follow out his solipsism; if he did, it would lead to a non-Russellian realism. A one-limit view self-destructs; we are not left, at the end of the *Tractatus*, with a philosophical view about a 'far side' of the 'limit', but merely with there being the sentences of our language. ... The Russell notion of the 'limit' of experience is meant to be the notion of something about which we can ask: 'Can we get beyond it, and if so how?' The *Tractatus* technique first makes available a criticism of the Russellian answer to that question; we are then meant to see that the Russell question has been shown not to be a question at all.[21]

James Conant explains the self-destruction of the idea of the limits of language as follows:

> as readers of the *Tractatus* ... we are drawn into the illusion of occupying a certain sort of a perspective. From this perspective, we take ourselves to be able to survey the possibilities which undergird how we must represent things as being, fixing what is 'logically' necessary and what is merely contingent. From this perspective, we contemplate the logical structure of thought as it is and imagine that we are also able to contemplate the possibility of its being otherwise. We take ourselves to be occupying a perspective from which we can view the logical structure of language 'from sideways on'.

However, we come to recognize that this perspective is illusory through a process of attempting to theoretically articulate it –a process which terminates in our loss of any assurance that we are making sense:

> The assumption underlying Tractarian elucidation is that the only way to free oneself from such illusions is to fully enter them and explore them from the inside. ... The illusion that the *Tractatus* seeks to explode, above all, is that we can run up against the limits of language. The book starts with a warning about a certain kind of enterprise – one of attempting to draw a limit to thought. In the body of the text, we are offered (what appears to be) a doctrine about 'the limits of thought'. With the aid of this doctrine we imagine ourselves to be able both to draw these limits and to see beyond them. We imagine ourselves able to do what the Preface warns we will fall into imagining ourselves able to do (once we imagine ourselves able to draw a limit to thought): we imagine ourselves able 'to think both sides of the limit' (and hence 'able to think what cannot be thought'). The aim of the work is to show us that beyond 'the limits of language' lies, not ineffable truth, but rather (as the preface cautions) *einfach Unsinn* [simply nonsense].[22]

In my opinion, Conant and Diamond are absolutely right to see the *Tractatus* as undermining the very project of trying to establish the limits of language, thought and world, rather than as trying to establish these limits through some sort of ineffable insight. If the question of the limits of thought is answerable only by something which cannot be *said*, then there is no answer, and there is no question either:

> For an answer which cannot be expressed the question too cannot be expressed.
> *The riddle* does not exist.
> If a question can be put at all, then it *can* also be answered. (TLP 6.5)

Therefore, the *Tractatus* rejects as illusory the very notion of a 'limit' of language or the world; but to do so is to reject as illusory the notion of the 'metaphysical subject' which *is* a 'limit' of the world. What has not been fully appreciated up to this point is the ethical significance of this result. In what follows I will try to make this clear.

I have suggested that, for Wittgenstein, solipsism is the beginning of a kind of a path of purification. That some such connection existed for him is clear from an entry that he made in his wartime diary after reading some of Nietzsche's works:

> I am strongly affected by his hostility against Christianity. Because his writings too have some truth in them. To be sure, Christianity is the only *sure* way to happiness; but what if someone spurned this happiness? Might it not be better to perish unhappily in the hopeless struggle against the external world? But such a life is senseless. But why not lead a senseless life? Is it ignoble? – How can it be reconciled with the strict solipsistic position (*streng solipsistischen Standpunkt*)? But what must I do in order that my life shall not be lost to me? I must be conscious of it always.[23]

Here the 'strict solipsistic position' seems to be functioning as some kind of test of the way of life Wittgenstein finds presented as an alternative to Christianity by Nietzsche. So in some sense, for Wittgenstein, solipsism, 'strictly carried out', (*streng durchgeführt*, 5.64) has a part to play in the way of life he would lead.

III

You may have been wondering for some time now in what sense this could be true. Before turning to a more detailed discussion of the text itself, I want to introduce a simple example which may help to set us thinking in the right direction. Leibniz is one philosopher whom we know Wittgenstein admired, and whose metaphysics has, in one sense, a quasi-solipsistic tinge.[24] For Leibniz, the spiritual substances, or monads, which make up the created world are isolated from one another, each representing the entire universe from its own perspective, yet each in causal relationship only to God (although all in pre-established harmony with one another). In the 'Discourse on Metaphysics', section 32, Leibniz tries to establish 'The Utility of These Principles in Matters of Piety and Religion'. He writes:

> We also see that every substance has a perfect spontaneity (which becomes freedom in intelligent substances), that everything that happens to it is a consequence of its idea or its being, and that nothing determines it, except God alone. And that is why a person of very exalted mind, revered for her saintliness, was in the habit of saying that the soul must often think as if there were nothing but God and itself in the world.[25]

The 'person of very exalted mind' whom Leibniz had in mind was the Spanish mystic St Teresa of Avila. In a letter to a mystically minded correspondent, Andreas Morell, he wrote:

> As to St. Teresa, you have reason to esteem her works, I found there one day this lovely thought that the soul must conceive of things as if there were only God and itself in the world. This even yields a considerable reflection in philosophy, which I usefully employed in one of my hypotheses.[26]

Leibniz's source was probably St Teresa's *Life*, where we read in chapter XIII: 'the utmost we have to do at first is to take care of our soul and to remember that in the entire world there is only God and the soul; and this is a thing which it is very profitable to remember'.[27]

In the context in which St Teresa makes this remark, her concern is to offer practical advice to 'beginners' in the practices of prayer and contemplation. Such beginners are subject to a number of temptations, one of which is to 'desire that everyone should be extremely spiritual when one is beginning to find what tranquillity, and what profit, spirituality brings'. This temptation is to be avoided because the beginner is not yet a good enough model of the spirituality she would seek to inculcate in others. Teresa speaks from her own experience:

> When... I tried to get others to practise prayer, and when on the one hand they would hear me saying so much about the blessedness of prayer, while on the other hand they would observe that I, who practised it, was so poverty-stricken in virtue, it would lead them into temptations and various kinds of foolishness.[28]

Seen in this light, Teresa's remark is like the advice not to try to remove a speck from your brother's eye when there is a beam in your own – her counsel is one of humility in spiritual life: 'Let us strive, then, always to look at the virtues and the good qualities which we find in others, and to keep our own grievous sins before our eyes so that we may be blind to their defects.'[29] Far from being solipsistic in any metaphysically serious sense, Teresa's counsel is designed to promote better relationships between the spiritual novice and her neighbors. In this light, Leibniz' appeal to her writings has to appear as a hollow manipulation in which he tries to extract the authority of her reputation for piety as a sanction for his elaborate metaphysical system.[30] Yet Leibniz' correspondent, Morell, saw a deeper meaning in another sense in St Teresa's saying: 'The thought that the soul is to consider the world as if there were nothing but God and itself in the world is founded on nothing but a veritable self-denial, and is excellent.'[31] St Teresa's disciple and fellow-worker St John of the Cross gave as a spiritual maxim: 'Live in this world as if there were in it but God and thy soul, so that thy heart may be detained by naught that is human.'[32] Here St John counsels an attitude of detachment from the world: 'Divest thyself of what is human in order to serve God.'[33] This is, of course, one of the great thoughts of the mystical tradition and the fount of ascetic practices, voluntary poverty and acts of mortification which we often find hard to understand today. Here we see the association of detachment from the world with the adoption of a quasi-solipsistic stance: act as if you are alone in the world (with God). Part of the key to understanding Wittgenstein's discussion of solipsism, I think, is to make this connection – but it is only *part* of the key.

Wittgenstein opens the discussion of solipsism with the pronouncement: '*The limits of my language* mean the limits of my world' (TLP 5.6). As several interpreters have pointed out, the unexpected intrusion of an essential use of the first-person possessive pronoun 'my' here shocks the reader, given the abstract and almost authorless discussion which makes up the *Tractatus* up to this point. I see the placement of this text and this pronoun at this particular juncture as quite deliberate.[34] For the argument of the *Tractatus* up to this point has been one concerted effort to *make language mine* through the construction of a theoretical apparatus adequate to enable me to *understand* language in the sense of establishing and bringing to explicit knowledge the essential nature of all representation. This effort to make language mine, to master and possess it, is at the same time an effort to master and possess the world, which since the very first pages of the book has been constrained to conform to the structures of language as a seeming condition of the possibility of meaningful language. Thus the opening sentence of the *Tractatus*, 'The world is the totality of facts, not of things' (TLP 1), imposes already on reality the structure of a proposition so that, if we can secure an understanding of the general *propositional* form, we will at the same time have grasped the essence of reality, the 'nature of all Being', as Wittgenstein puts it in his wartime *Notebooks*: 'My *whole* task consists in explaining the nature of the proposition. That is to say, in giving the nature of all facts, whose picture the proposition *is*. In giving the nature of all Being.'[35]

The significance of this way of looking at the project of the *Tractatus* up to TLP 5.6 ('*The limits of my language* mean the limits of my world') can be brought out by considering some of Wittgenstein's worries about the nature of the will and of the purpose of life in the *Notebooks*. These topics make a sudden appearance in the midst of purely logical discussions in the entry for June 11, 1916: 'What do I know about God and the purpose of life?' There follows a long series of remarks, ending with: 'I cannot bend the happenings of the world to my will: I am completely powerless. I can only make myself independent of the world – and so in a certain sense master it – by renouncing any influence on happenings' (NB 72-73).

Much of the remainder of the *Notebooks*, through January, 1917, is taken up with the topics introduced here; we will discuss this further below. For the moment I want to highlight one entry from this long and complex discussion, written two months later, on 8 August 1916:

> How can man be happy at all, since he cannot ward off the misery of this world?
> Through the life of knowledge.
> The good conscience is the happiness that the life of knowledge preserves.
> The life of knowledge is the life that is happy in spite of the misery of the world.
> The only life that is happy is the life that can renounce the amenities of the world.
> To it the amenities of the world are so many graces of fate. (NB 81)

These two *Notebooks* entries evince a desire to solve the problem of the meaning of life through a form of renunciation of the world, made possible by a 'life of knowledge'. I suggest that the knowledge in question is the very knowledge that Wittgenstein took himself to be seeking in trying to give 'the nature of all Being' – the knowledge of the *limits* of language and the world. In giving these limits, he would be establishing once and for all the necessary structures of the world; he would therefore also be establishing the set of all possibilities. This determination of the limits of possibility would allow him to master the world, by adopting an attitude of indifference towards the *happenings* of the world – those states-of-affairs within the strictly delimited set of all possible states-of-affairs which just *happen* to obtain.

IV

So the author of the *Tractatus,* seems to himself to have followed the path of knowledge to achieve a mastery over language and the world; thus both language and world have become, for him, *mine*. (The same can be said for the reader of the work; in working through the text one is led down the same path that Wittgenstein followed. Thus reading the book is an imitative act, and insofar as the writing of the book is an 'ethical deed' so is its reading.) Thus he thinks that he can go on to say: 'That the world is *my* world, shows itself in the fact that the limits of the language (*the* language which I understand) mean the limits of my world' (TLP 5.62).[36] '*The* language which I understand' is the language which I have *made* mine through coming to *know* how it works, how it manages to represent the world, which is therefore also *mine*. I become all-embracing, all-possessing – 'The world and life are one. I am my world. (The microcosm.)' (TLP 5.621-5.63.)

Yet this 'achievement' of mastery over the world begins to unravel before it is even finished being proclaimed. For 5.61 reminds us that, according to the theory of language as pictorial representation developed in the earlier part of the book, the limits of language and the world are also the limits of *logic* and so of *thought*. Logic, language and thought cannot 'get outside' these limits, for that would be, as the Preface warned us, only to produce nonsense. Yet if we can say 'there is the I' then this 'I' cannot be *in* the world, for 'we cannot ... say in logic: This and this there is in the world, that there is not'. Thus when we deploy the terminology of 'showing' in 5.62 to claim that the world's being *mine* 'shows itself' in 'the fact that the limits of the language ... mean the limits of *my* world' we are driven into an irresolute reading of 'showing' – into the nonsense of claiming that something shows itself which cannot be said – and yet, it seems we *do* say it – 'I am my world'.

Wittgenstein develops this problem in a number of different ways, all of which lead to the conclusion that the 'I' which tries to claim the world as 'mine' through mastering it is not *in* the world at all. At 5.631 we are given the image of a book titled 'The world as I found it'. This book would purport to be a description of *my* world. In this book, he claims, there will be mention of my body, and of which parts of the world obey my will; but in this book there will be *no* mention of the 'I' of the title. Thus, 'in an important sense there is no subject'. But of course there is mention of 'I' – not *in* the book, but in its title. At 5.633 we are given another image, that of the eye and the visual field. As the eye is not found *within* the visual field, so the 'I' is not found *within* the world. But of course the eye *is* somewhere, only not somewhere in my visual space, but rather somewhere in physical space.

These metaphors suggest that the 'I' of solipsism must be located not *in* the world but somewhere 'outside' of it. But there is no 'outside' of the world, the totality of facts. So the 'I' must be located not quite 'outside' while yet not being 'inside' – at the 'limit' of the world.

Peter Sullivan has argued persuasively that these arguments against the presence of the self of solipsism *in* the world are grounded in Wittgenstein's rejection of a substantial *a priori*: 'There is no order of things a priori' (TLP 5.634).[37] If the eye were a part of the visual field, this would provide an *a priori* order to our visual experience, privileging one part of that experience over others. If the self of solipsism were a part of the world, then the world would have a privileged centre, a fixed point of origin for the coordinates of logical space, as it were. For the self of solipsism, by acquiring knowledge of the necessary conditions of the possibility of language and the world, has become one of those necessary conditions. If the self were a part of the world, then facts about that self – facts recording its possession of knowledge of the structure of language, thought and reality – would have the status of *a priori* truths. Thought would have gotten beyond the limits of logic. Thus, by mastering the world, the self has written itself out of that world.

The consequence of this is what David Bell calls 'self-effacing solipsism':[38] 'Here we see that solipsism strictly carried out coincides with pure realism. The I in solipsism shrinks to an extensionless point and there remains the reality co-ordinated with it' (TLP 5.64). Yet one might argue that the self has not been effaced here, but merely displaced to the 'limit' of the world from whence it remains the essential defining condition of the world.[39] For does not Wittgenstein hasten to add:

> There is therefore really a sense in which in philosophy we can talk of a non-psychological I.
> The I occurs in philosophy through the fact that the 'world is my world.'
> The philosophical I is not the man, not the human body or the human soul of which psychology treats, but the metaphysical subject, the limit – not a part of the world.

Here, we might say, the philosophical self is not the 'I' of the title of 'The World as I found it' – for the title of the book can be dispensed with – but its author, without which the book does not exist. It is not the physical eye – which, from within the visual field, cannot be proved to exist – but the 'geometrical eye' which structures that field internally. From this point of view, it might appear that in being displaced to the limit of the world, the self of solipsism, far from having shrunk to an extensionless point, has come to occupy the place of greatest importance. *It is* the limit, the author, the structurer, the ground of all Being.

To take this line, however, one must take seriously the talk of the philosophical self as a 'limit' of the world. I have already indicated my agreement with those interpreters who see the *Tractatus* rather as trying to destabilize the very idea of a 'limit' at work here. What I want to show now is that the role that we take this idea to play in the *Tractatus* will have a decisive influence on our reading of the ethical significance of the book. In order to see this, we need to discuss Wittgenstein's conception of the *will* in the *Notebooks* and the *Tractatus*.

V

As we saw, Wittgenstein's first recorded reflections on the purpose of life are connected to the problem of the will: 'I cannot bend the happenings of the world to my will: I am completely powerless' (NB 73). 'My will' here is identified with my *desires*, wants and wishes. Classically, these desires are seen as the *cause* of my *actions*. But Wittgenstein maintains that the atomic facts which make up the world are logically independent. Therefore 'The events of the future *cannot* be inferred from those of the present. The belief in the causal nexus is superstition.' (TLP 5.1361) and again 'A necessity for one thing to happen because another has happened does not exist. There is only *logical* necessity' (TLP 6.37). In consequence, the will is free, but powerless: 'The freedom of the will consists in the fact that future actions cannot be known now. We could only know them if causality were an *inner* necessity, like that of logical deduction' (TLP 5.1362). 'The world is independent of my will. Even if everything we wished were to happen, this would only be, so to speak, a favour of fate, for there is no *logical* connexion between will and world, which would guarantee this, and the assumed physical connexion itself we could not again will' (TLP 6.373-6.374). These last two remarks are repeated *verbatim* from the *Notebooks* entry for 5 July 1916, the day after Wittgenstein recorded his first thoughts on God and the purpose of life.

On that same day Wittgenstein notes an alternative way in which to think of the 'will' as related to the world not from inside as the cause of action, but from outside, from the limit:

> If good or evil willing affects the world it can only affect the boundaries of the world, not the facts, what cannot be portrayed by language but can only be shewn in language.
> In short, it must make the world a wholly different one.
> The world must, so to speak, wax or wane as a whole. As if by accession or loss of meaning. (NB 73)

The *Tractatus* repeats these thoughts, in almost the same words, at 6.43, and adds: 'The world of the happy is quite another than that of the unhappy'.[40]

But how can will 'affect the boundaries of the world' in this way? Wittgenstein is drawn to the following kinds of answer:

> The world is *given* me, i.e. my will enters into the world completely from outside as into something that is already there.
> Good and evil enter only through the *subject*. And the subject is not part of the world, but a limit of the world. (NB 74)

> ...the willing subject would have to be happy or unhappy, and happiness and

> unhappiness could not be a part of the world.
> As the subject is not a part of the world but a presupposition of its existence, so good and evil which are predicates of the subject, are not properties in the world. (NB 79)

In short, the will can affect the limits of the world because the willing subject simply *is* that limit. 'The will is an attitude of the subject towards the world. The subject is the willing subject' (NB 87). My unhappiness stems from a lack of co-ordination between my antecedent *wants* and the contingent facts which I find in the world; I can then make myself happy by *renouncing* all such wants, counting them as nothing, and adopting an *attitude* of acceptance towards whatever happens.

> I can only make myself independent of the world – and so in a certain sense master it – by renouncing any influence on happenings.
> ... we are in a certain sense dependent, and what we are dependent on we can call God.
> In this sense God would simply be fate, or, what is the same thing: The world – which is independent of my will.
> I can make myself independent of fate.
> There are two godheads: the world and my independent I.
> I am either happy or unhappy, that is all. It can be said: good or evil do not exist.
> In order to live happily I must be in agreement with the world. And that is what 'being happy' *means*.
> I am then, so to speak, in agreement with that alien will on which I appear dependent. That is to say: 'I am doing the will of God'.
> Live happy! (NB 73-75)

In 'living happy' I make myself independent of fate – independent of God. By 'doing the will of God' – agreeing with whatever the world brings me – I become a godhead – 'the independent I'.

Wittgenstein would have been familiar with this sort of solipsistic response to the problem of life from his youthful reading of Otto Weininger's *Sex and Character*.[41] Here is Weininger's account of the point of Kant's ethics:

> Having made this determination, we now understand this 'Critique of Practical Reason' as follows. Man is alone in the universe, in eternal monstrous loneliness.
>
> He has no purpose outside of himself, he lives for nothing else – he has flown away, far beyond wanting-to-be-a-slave, being-able-to-be-a-slave, having-to-be-a-slave: deep under him, all human society has disappeared, social ethics has sunk; he is alone, *alone*.
>
> But he has now become one and all; and because of that he also has a law in himself, because of that he is himself all law, and no capricious arbitrary will. And he asks of himself, that he should follow this law in himself, the law of his self, that he should be only law, without hindsight or deference (*Rück-Sicht hinter sich*), without foresight or caution (*Vor-Sicht vor sich*). This is the dreadful greatness: that he obeys duty has no *further* sense. Nothing is superior to him, the alone, the all-one. (*Nichts ist ihm, dem Alleinen, All-Einen übergeordnet.*) ... Kant's most lonely man does not laugh and does not dance, he does not bawl and does not rejoice: he has no need to make a noise, since the universe is too deeply silent. The senselessness (*Sinnlosigkeit*) of a groundless world (*einer Welt 'von ohngefähr'*) is not duty to him; rather, to him his duty is the

> sense of the universe. Saying yes to *this* loneliness, that is the 'Dionysian' moment in Kant: only that is morality.[42]

If Wittgenstein's view in the *Tractatus* were anything like this, one could appreciate why one might see 'selflessness' as 'both an ambition and a thwarted one at that in the early thought of Wittgenstein', as Emyr Thomas puts it.[43] One could understand James Edwards' assessment: 'the outcome of the *Tractatus* ... is narcissistic: the self is the maker of meaning'.[44] Edwards sees the ethical vision of the *Tractatus* in terms of a heroic struggle towards self-perfection, a 'transfiguration' of the self through 'ascent to a godlike self-consciousness'. This self-consciousness is also the 'contemplation of the world *sub specie aeterni*', the 'feeling of the world as a limited whole' which Wittgenstein identifies with '*das Mystische*' at 6.45. But here, as Edwards remarks, we have not self-denial, but 'a manifestation of heroic self-assertion ... the truest exemplification of the will-to-power. And, in the end, once the ladder is kicked away, only naked will remains'.[45]

VI

Yet for anyone familiar with Wittgenstein's writings of the time, particularly his correspondence with his friend Paul Engelmann, this description of his ethical vision must strike a discordant note. It neglects an influence as important for the early Wittgenstein as Schopenhauer and Weininger: his reading of such Christian and quasi-Christian writers as Kierkegaard, Tolstoy, Dostoyevsky[46] and the seventeenth century Silesian mystic Johann Scheffler (also known as Angelus Silesius). Wittgenstein was profoundly influenced by Tolstoy's rendering of the message of Jesus, *The Gospel in Brief*, which he discovered during the war. 'At its time', he told von Ficker, 'this book virtually kept me alive.'[47] Yet Tolstoy preached not self-assertion but humble service to others and love of one's neighbour. Wittgenstein's own actions after the end of the First World War, giving up philosophy and taking up a career as a schoolteacher in rural Austria, show the seriousness with which he took this teaching.

Edwards tries to accommodate this dimension of Wittgenstein's thought and life and explicitly recognizes Tolstoy's influence. He discusses a passage from the *Notebooks* in which Wittgenstein writes:

> Is only he happy who does *not* will?
> 'To love one's neighbour' means to will!
> But can one want and yet not be unhappy if the want does not attain fulfillment? (And this possibility always exists.)
> Is it according to the common conceptions, good to want *nothing* for one's neighbour, neither good nor evil?
> And yet in a certain sense it seems that not wanting is the only good.
> Here I am still making crude mistakes! No doubt of that! (NB 77-78)

Edwards comments that this passage:

> shows both [Wittgenstein's] temptations and his resistance. ... One can see Wittgenstein tempted to identify the good life with the renunciation of desire ... a Schopenhauerian influence; yet he never succumbs. ... The Christian influence of Tolstoy, calling him to a life of active love of the least fortunate of men, is stronger than the advice of Schopenhauer to turn away from wanting any alteration in the necessary fabric of things.[48]

Yet the characterization of the self and the will as limits of the world which Edwards ultimately attributes to Wittgenstein seems to leave no room for a 'life of active love' unless 'to love one's neighbour as oneself' is understood as 'affirming whatever happens in the life of one's neighbour in the act of affirming the world as my world'.

Similarly, Edwards' attempts to account for Wittgenstein's own life choices on completing the *Tractatus* fall flat. Edwards tells us that:

> in going to the mountains to teach in the peasant schools, there is ... a corresponding *practical* attempt heroically to ascend from the normal human condition and to put on the attributes of a god. The image that springs immediately to mind is the *imitatio Christi*. Not only does his life in the mountains reflect that service to the poor and humble which Jesus (and Tolstoy) preached; it represents Wittgenstein's own self-conscious attempt to become Spirit, to become a god.[49]

Yet to truly imitate Christ is not to serve the poor and humble *in order to* become a god. It is to be poor and humble in the service of others. To imitate Christ one must climb not the mountain of the transfiguration but Golgotha. As St Paul puts it:

> Your attitude should be the same as that of Christ Jesus: Who, being in very nature God, did not consider equality with God something to be grasped, but emptied himself, taking the very nature of a slave, being made in human likeness. And being found in appearance as a man, he humbled himself, and became obedient to death – even death on a cross! Therefore God exalted him to the highest place and gave him the name that is above every other name, that at the name of Jesus every knee should bow, in heaven and on earth and under the earth, and every tongue confess that Jesus Christ is Lord, to the glory of God the Father.[50]

The imitation of Christ is found in that self-denial which not only renounces the desire for all the amenities of the world, but renounces the desire to be a god, to *grasp*, to *possess* 'equality with God'. It is this self-denial which leads to exaltation – not exaltation sought as a form of self-aggrandizement, however, but exaltation found in the radical undermining and transformation of the very nature of our wanting and willing.[51]

By the time of writing the *Tractatus,* Wittgenstein would have encountered such thoughts in the mystical couplets of Angelus Silesius' *Cherubinic Wanderer*.[52]

Silesius preaches detachment (*Gelassenheit*):

> I, 22 Detachment
> As much as you abandon to God, so much may He be for you,
> Not less and not more will He help you out of your troubles.[53]
>
> I, 44 One must let go of the something
> Man, if you love something, you don't truly love;
> God is not this and that, so let the something go completely.[54]

But this detachment is incomplete unless it culminates in the extinction of will, and the negation of the 'I':

> VI, 191 To renounce the world is to renounce little
> The whole world is nothing; You have despised very little,
> Even if you have brought the world out of your Spirit.
>
> VI, 192 To renounce yourself is to renounce something
> You must get out of yourself. If you hate your own self,
> then I will take you to have renounced something.
>
> VI, 193 One must be killed
> Everything must be slaughtered. If you don't slaughter yourself for God,
> Eternal death in the end will slaughter you for the enemy.[55]

Yet this teaching can seem mysterious, since Silesius also teaches that we are to do God's will:

> I, 281: His commandments are not difficult
> Man, if you live in God, and die to your will,
> then nothing is so easy for you, as to fulfill God's commands.[56]

How, one might ask, can death to self and will allow me to fulfill God's commandments – especially the commandment 'to love your neighbor as yourself'? As Wittgenstein exclaimed in the *Notebooks*, 'to love one's neighbor is to will!' and to love one's neighbor as one's self, one must have a self. The answer is that we are to empty ourselves so that God may fill us:

> I, 138 The more you go out, the more God enters
> The more you can pour yourself out of yourself,
> the more must God with his Godhead flow into you.[57]
>
> II, 136 Detachment
> Go out, God goes in: die to yourself, God lives in you:
> Be not, it is He: do nothing, the command is fulfilled.[58]

Our will, then, becomes truly one with God's not through our adopting a stance in which we accept whatever occurs, but rather through our opening ourselves to serve as channels of God's life and activity. As William James, explains in *The Varieties of Religious Experience*, in a passage which Wittgenstein surely knew well: 'In Paul's language, I live, yet not I, but Christ liveth in me. Only when I become as nothing can God enter in and no difference between his life and mine remain outstanding'.[59] Again, here is Angelus Silesius:

> II, 144 What is detachment?
> What is detachment? I say without hypocrisy:
> That it is the will of Jesus in your soul.[60]

Seen in this light, the Tractarian conclusion that 'solipsism strictly carried out coincides with pure realism. The I in solipsism shrinks to an extensionless point and there remains the reality co-ordinated with it' (TLP 5.64) takes on a different appearance. It appears as the very model of self-emptying, of that denial of self which is needed before the divine life can enter in. But what of the 'metaphysical self' of which, according to TLP 5.641, philosophy can speak? Even this self, I maintain, has to be killed. As Simone Weil puts it:

> We possess nothing in the world – a mere chance can strip us of everything – except the power to say 'I'. That is what we have to give to God – in other words, to destroy. There is absolutely no other free act which it is given to us to accomplish – only the destruction of the 'I'.[61]

Wittgenstein's analogy of the 'I' of solipsism with the eye of the field of vision can help us to see why it is necessary to destroy this 'I'. As Wittgenstein was later to point out, my visual field 'has no neighbours'. It is equally so for the 'I' of solipsism. For the metaphysical self, the limit of the world, other people can appear only as objects *in* the world, not as other selves. But to love one's neighbour as one's self, one must have neighbours! So to love one's neighbour as one's self, one must kill the 'I' that recognizes no neighbours – the metaphysical self. Of course, this is not to kill any person, but only an illusion, although an illusion which is sustained and cherished because it answers to a desire – the desire to master the world, to be God.

VII

I stated above my agreement with Diamond, Conant and others that the *Tractatus* aims in the end not to trace out the limits of language, thought and reality, but to explode the illusion that there are such limits to be drawn. Now we can see the full ethical significance of this act – to explode this illusion is to explode a cherished myth, because it is to explode the illusion of the godhead of the independent 'I'.

Solipsism is, in part, a truth, *because* it leads to its own elimination. Solipsism is, strictly followed through, the self-humbling of pride.[62]

It is true that Wittgenstein tells us that there is 'really a sense in which in philosophy we can talk of a non-psychological I'. But we must recall that 'philosophy' for Wittgenstein is 'not a theory but an activity' whose object is 'the logical clarification of thoughts' (TLP 4.112). Philosophical problems are to be unmasked as nonsense through revealing their sources in 'the misunderstanding of the logic of our language', (Preface) so that their solutions will be found 'in the vanishing of the problem' (TLP 6.521). Wittgenstein tells us that: 'The I occurs in philosophy through the fact that the "world is my world"' (TLP 5.641), but note well the quotation marks around the phrase 'the world is my world'. Wittgenstein is signalling here, I think, that there is no real 'fact' that the world is my world (the words 'the fact that' do not occur in the German text). There is only a bit of language – 'the world is my world' – which we ultimately recognize to be nonsense. But in recognizing this to be nonsense, we do not only give up a philosophical misconception, we give up at the same time a conception of ourselves as 'alone in the universe, in eternal monstrous loneliness' and a life-project grounded on that conception, 'saying yes to *this* loneliness, the "Dionysian" moment in Kant'.

But what, you might ask, does this leave us with? What is our conception of ourselves, and what projects are we left to pursue? Having disposed of the 'I' of solipsism, am I left with any 'self' at all? Having renounced not only the amenities of the world, but my will as well, how will I act? As we saw above, to love one's neighbour as one's self, one needs a self – and one must will.

To try to answer these questions in terms of some new theory of the self and the will would of course do violence to Wittgenstein's conception of philosophy as 'not a theory, but an activity'. Still, I think, something *can* be said in response to them. At the end of the *Tractatus*, we are left with *ourselves*. We are left, in particular, with Ludwig Wittgenstein, the author of this book, and his readers – human beings, capable of communicating with, and understanding, one another. 'My propositions are elucidatory in this way: the one who understands me eventually recognizes them as nonsense' (TLP 6.54). 'My' and 'me' here are not nonsensical attempts to express the mysterious 'I' of solipsism. In the *Notebooks*, Wittgenstein exclaims 'The I, the I is what is deeply mysterious!' (NB 80). To come to see this as a philosophical problem is to come to see it as resting on 'the misunderstanding of the logic of our language', and so to see it vanish.

Not only are we left with ourselves, but we are enabled to simply *act* in the world. In the *Notebooks*, as we have seen, Wittgenstein rejected a conception of the will as the underlying *cause* of our actions, and was drawn to a conception of the will as adopting an attitude or stance towards the world as a whole. He suggested there that to 'be happy' it was only necessary to 'be in agreement with the world' and that this could then be called 'doing the will of God'. (NB 75). But now we see that *this* way of 'doing the will of God' is just a self-seeking attempt to secure personal happiness. To truly do the will of God is simply to *act*, to do as

God commands, so that 'not I, but God liveth in me'. Here we find yet another conception of the will that Wittgenstein at least briefly explored in the *Notebooks*: 'The act of the will is not the cause of the action but is the action itself' (NB 87). Of course, this is not really a 'conception' of the will at all, in the sense of a theoretically articulated account. But that is not something we should ask of *Wittgenstein*, if we take the spirit of his philosophy at all seriously.

So, I have argued, at the end of the *Tractatus*, when we have thrown away the ladder, we are simply left with ourselves, acting in the world. But, one might object that this leaves no room for 'the mystical ways of thinking' in which Russell had found Wittgenstein absorbed. In one way, I think this is right and in another way wrong. For, as his friend Paul Engelmann tells us, Wittgenstein 'was never a mystic in the sense of occupying his mind with mystic-gnostic fantasies. Nothing was further from his mind than the attempt to paint a picture of a world beyond ... about which we cannot speak'.[63] Readings of the *Tractatus*' mysticism commonly rely on 6.45: 'The feeling of the world as a limited whole is the mystical feeling'. But, on my reading, the concluding sections of the book, the 6.5s, proclaim the end of all such conceptions of 'limits' of the world, exposing them as 'simple nonsense'. Nonetheless, 6.522 asserts that 'There is indeed the inexpressible. This *shows* itself; it is the mystical'. Here we have an occurrence of 'showing' which, I think, can be redeemed through being given a *practical* reading. When the mystical shows itself, one 'sees the world rightly' – one knows how to live. The connection of this thought to the properly mystical tradition is made clear in the following saying of St John of the Cross:

> The venerable and blessed father Fray John of the Cross was once asked how a man went into ecstasy. 'By renouncing his own will,' he replied, 'and doing the will of God. For ecstasy is naught but the going forth of a soul from itself and its being caught up in God, and this is what happens to the soul that is obedient, namely, that it goes forth from itself and from its own desires, and, thus lightened, becomes immersed in God.'[64]

When we climb the ladder of the *Tractatus*, we are 'going forth from ourselves'. For the ladder is not only a ladder of philosophical nonsense, but a ladder of our own desires, and of the illusions they enmesh us in. When we throw the ladder away, we are lightened of our burden, and enabled to simply do the will of God. We overcome (*überwinden*) not only Wittgenstein's propositions, but ourselves. In St Paul's words, we 'put off [the] old self, which is being corrupted by its deceitful desires' and 'put on the new self, created to be like God in true righteousness and holiness'.[65] Thus 'made new' we will live *in* the world, in service of others, without being *of* the world.

At least, this is how things are supposed to work – but Wittgenstein's own life experience, his failure to truly serve the students to whom he was assigned as a teacher, showed him that salvation is not to be had as cheaply as he had hoped. That, however, is another story.

Notes

1 Versions of this paper were presented to the philosophy departments of the University of Notre Dame and the University of East Anglia. I thank the members of both audiences for helpful discussions. Thanks are especially due to David Burrell for comments on an earlier version of this paper.
2 Ludwig Wittgenstein, *Tractatus Logico-Philosophicus*, trans. C.K. Ogden (London: Routledge, 1922) hereafter referred to as 'TLP' in the text. Also translated by B.F. McGuinness and D.F. Pears in 1961. In general I will follow Ogden's translation of the *Tractatus*, as I do here. Occasionally (as in rendering '*unsinnig*' by 'nonsense' in TLP 6.54 below) I will follow the Pears-McGuinness translation, or combine the two, incorporating modifications of my own.
3 See for example P.M.S. Hacker, *Insight and Illusion: Themes in the Philosophy of Wittgenstein* (Oxford: Clarendon Press, 1986), ch. IV; also David Pears, *The False Prison: A Study of the Development of Wittgenstein's Philosophy*, Vol. 1 (Oxford: Clarendon Press, 1987), ch. 7; Juliet Floyd, 'The Uncaptive Eye: Solipsism in Wittgenstein's *Tractatus*', in L.S. Rouner (ed.), *Loneliness*, Boston University Studies in Philosophy and Religion (Notre Dame: University of Notre Dame Press, 1998); David Pears, 'The Originality of Wittgenstein's Investigation of Solipsism', *European Journal of Philosophy*, **4**, 1996, pp. 124-136; David Bell, 'Solipsism and Self-Reference', *European Journal of Philosophy*, **4**, 1996, pp. 155-174; Peter M. Sullivan, 'The "Truth" in Solipsism and Wittgenstein's Rejection of the A Priori', *European Journal of Philosophy*, **4**, 1996, pp. 195-219; H.O. Mounce, 'Philosophy, Solipsism and Thought', *Philosophical Quarterly*, **47**, 1997, pp. 1-18; and Cora Diamond, 'Does Bismarck Have a Beetle in his Box? The Private Language Argument in the *Tractatus*', in A. Crary and R. Read (eds), *The New Wittgenstein*, (London: Routledge, 2000).

Hacker's interpretation brings in ethical themes, in a way more fully developed by James C. Edwards (see note 6); Floyd's essay also addresses these themes in a way to which I am more sympathetic, yet without working out the issues explored in this paper.

4 Brian McGuinness, *Wittgenstein: A Life; Young Ludwig 1889-1921* (Berkeley: The Univeristy of California Press, 1988), p. 228.
5 Simone Weil, *Gravity and Grace* (London and New York: Routledge, 1963), pp. 56-7. The initial inspiration for much of my reading of Wittgenstein on solipsism came from reading Weil on detachment and self-abandonment.
6 In a well-known letter to Ludwig Ficker, Wittgenstein stated that 'the point of the book is ethical': C.G. Luckhardt (ed.), *Wittgenstein: Sources and Perspectives* (Ithaca, NY: Cornell University Press, 1979), p. 94. I provide an interpretation of this claim in my 'The Purpose of *Tractarian* Nonsense', *Noûs*, **35** (2001), pp. 39-73.
7 James C. Edwards, *Ethics without Philosophy: Wittgenstein and the Moral Life* (Tampa: University Presses of Florida, 1982); Emyr Vaughan Thomas, 'From Detachment to Immersion: Wittgenstein and the "Problem of Life"', *Ratio*, **12**, (1999), pp. 195-209. Thomas' interpretation largely repeats (albeit unwittingly) Edwards' more developed reading, so in what follows I will concentrate on Edwards. (In fairness to Edwards, I should mention that, in private correspondence, he has indicated that he would not now want to be committed to all aspects of the interpretation developed in *Ethics without Philosophy*.)

8 Both Thomas and Edwards also read the later Wittgenstein as moving beyond the self-absorption that they find in the *Tractatus*. My interpretation, which sees this move as already present in the early work, thus emphasizes the continuity of Wittgenstein's ethical concern.

9 Quoted in B. McGuinness and G.H. von Wright (eds), *Ludwig Wittgenstein: Cambridge Letters*, (Oxford: Blackwell Publishers, 1995), p. 140.

10 Cora Diamond, 'Throwing Away the Ladder: How to Read the *Tractatus*', in *The Realistic Spirit: Wittgenstein, Philosophy and the Mind* (Cambridge, MA: MIT Press, 1991); James Conant, 'The Method of the *Tractatus*', in E.H. Reck (ed.), *From Frege to Wittgenstein: Perspectives on Early Analytic Philosophy* (Oxford: Oxford University Press, 2001).

11 Hacker, *Insight and Illusion*, *op cit.*, n. 3, pp. 99-100.

12 Ibid., p. 101.

13 This line of thought is expressed and explored by Edwards in *Ethics Without Philosophy*, *op. cit.*, n. 7. As we shall see, Edwards does not consistently follow through on his insight, falling into an irresolute reading of talk of the subject as the 'limit' of the world which 'shows itself'. This difficulty prevents him from seeing the fully developed significance of the way of purification which Wittgenstein is attempting to communicate in the *Tractatus*.

14 In this paragraph I borrow from my 'The Purpose of *Tractarian* Nonsense' *op. cit.*, n. 6, as well as my 'Mathematics and Meaning in the *Tractatus*', *Philosophical Investigations*, **25**, 2002, pp. 272-303.

15 I would like to call this an ambiguity, but that would be misleading since on one side we have nothing but nonsense.

16 John 3:21. The New International Version paraphrases this as 'whoever lives by the truth, comes into the light…'. See K. Barker (ed.), *The NIV Study Bible* (Grand Rapids: Zondervan, 1985), p. 1598. The King James translation 'But he that doeth truth, cometh into the light…' is more accurate here.

17 Again, I borrow from 'The Purpose of *Tractarian* Nonsense' *op. cit.*, n. 6, here.

18 According to the instructions provided at the beginning of the book, 5.62 comments not on 5.61 but on 5.6. Further confirmation of this reading can be gleaned from earlier versions of these remarks in the *Notebooks* and the *Prototractatus*, as is demonstrated by Peter Sullivan in 'The "Truth" in Solipsism', *op. cit.*, n. 3.

19 Hacker, *Insight and Illusion*, *op. cit.*, n. 3, p. 100.

20 Floyd, 'The Uncaptive Eye', *op. cit.*, n. 3, p. 80.

21 Diamond, 'Does Bismarck Have a Beetle in his Box?', *op. cit.*, n. 3, pp. 282-283; see also Conant, 'The Method of the *Tractatus*', *op. cit.*, n. 10, on which I draw in the next paragraph.

22 Conant, 'The Method of the *Tractatus*', , *op. cit.*, n. 10, pp. 422-424.

23 Diary entry for 8 December 1914, quoted in McGuinness, *Wittgenstein: A Life*, *op. cit.*, n. 4, p. 225. For the German, see Ludwig Wittgenstein, *Geheime Tägebucher 1914-1916*, ed. W. Baum (Vienna: Turia & Kant, 1991), p. 50.

24 When his friend and student M. O'C. Drury told Wittgenstein that he had to read Leibniz for his exams, Wittgenstein replied: 'Count yourself lucky to have so much time to study such a great man as Leibniz': M. O'C. Drury, 'Conversations with Wittgenstein', in R. Rhees (ed.), *Recollections of Wittgenstein* (Oxford: Oxford University Press, 1984), p. 105.

25 G.W. Leibniz, *Philosophical Essays*, ed. and trans. R. Ariew and D. Garber

(Indianapolis and Cambridge: Hackett Publishing, 1989), p. 64. A similar remark is made in idem, 'A New System of the Nature and Communication of Substances, and of the Union of the Soul and Body', in ibid., p. 143.

26 G.W. Leibniz, *Textes Inédits*, Vol. 1, ed. G. Grua (Paris: Presses Universitaires de France, 1948), p. 103 (my translation). Reference taken from Leibniz, *Philosophical Essays*, *op. cit.*, n. 25.

27 Teresa of Avila, *The Life of Teresa of Jesus: The Autobiography of Teresa of Avila*, ed. and trans. E. A. Peers (New York: Image Books, 1991), p. 142. Reference taken from G.W. Leibniz, *Discours de Métaphysique et Correspondance avec Arnauld*, ed. G. LeRoy, (Paris: Librairie Philosophique J. Vrin, 1984), p. 265.

28 Teresa, *Life*, *op. cit.*, n. 27, p. 141.

29 Ibid., p. 142.

30 As argued by Stuart Brown, 'Some Occult Influences on Leibniz's Monadology', in A.P. Coudert *et al.* (eds), *Leibniz, Mysticism and Religion* (Dordrecht: Kluwer, 1998), pp. 5-6.

31 Leibniz, *Textes Inédits*, *op.cit.*, n. 26, p. 109 (my translation).

32 John of the Cross, 'Points of Love', in *The Complete Works of Saint John of the Cross*, Volume III, ed. E.A. Peers, trans. P. S. de Santa Teresa (London: Burns, Oates and Washbourne, 1935), p. 256.

33 Ibid., p. 255.

34 I also see it as an important improvement over its placement in the *Prototractatus* manuscript, where the discussion of solipsism (TLP 5.6-5.641, *Prototractatus* 5.335-5.33552) occurs before the discussion of the forms of elementary propositions (TLP 5.55-5.5571, *Prototractatus* 5.34-5.4223) and also before the discussion of propositions of the form '*A* thinks that *p*' (TLP 5.54-5.5423, *Prototractatus* 6.001-6.0051), and is treated as in some way part of the commentary on Wittgenstein's treatment of the identity sign (*Prototractatus* 5.33).

35 Ludwig Wittgenstein, *Notebooks 1914-1916*, ed. G.H. von Wright and G.E.M. Anscombe, trans. G.E.M. Anscombe, 2nd edn (Chicago: University of Chicago Press, 1979), p. 39, hereafter referred to in the text as 'NB'. The entire apparently ontological development of the early 2s in the *Tractatus* is driven by linguistic considerations.

36 The phrase '*the* language which I understand' translates 'der *Sprache, die allein ich verstehe*', and there was at one time uncertainty whether this was to be rendered 'the language which only I understand' (which suggests Wittgenstein's adherence to a form of private language, of the sort he later criticized in the *Philosophical Investigations*) or 'the only language which I understand'. Ogden's translation, used here, and accepted by Wittgenstein, favours the second alternative, and this is confirmed by a notation made by Wittgenstein in Frank Ramsey's copy of the *Tractatus*. See Jaakko Hintikka, 'On Wittgenstein's "Solipsism"', *Mind*, **67**, 1958; and C. Lewy 'A Note on the Text of the *Tractatus*', *Mind*, **76**, 1967.

37 Sullivan, 'The 'Truth' in Solipsism', *op. cit.*, n. 3.

38 Bell, 'Solipsism and Self-Reference', *op. cit.*, n. 3, p. 160.

39 This is the conclusion reached in Thomas, 'From Detachment to Immersion', *op. cit.*, n. 7, p. 202.

40 This also repeats a thought recorded in Wittgesnstein's *Notebooks*, *op. cit.*, n. 35, p. 78, on 29 July 1916.

41 See Ray Monk, *Ludwig Wittgenstein: The Duty of Genius* (New York: Free Press, 1990), pp. 19ff. Monk sees Weininger's solipsism in the *Tractatus* (p. 190) and uses these lines from *Sex and Character* as the epigraph of his biography of Wittgenstein: 'Logic and ethics are fundamentally the same, they are no more than duty to oneself.'
42 Otto Weininger, *Geschlecht und Character* (Munich: Mathes & Setz Verlag, 1980), pp. 210-211 (my translation). (I thank Marian David and Vittorio Hösle for help with the translation of this passage.)
43 Thomas, 'From Detachment to Immersion', *op. cit.*, n. 7, p. 202.
44 Edwards, *Ethics without Philosophy, op. cit.*, n. 7, p. 71.
45 Ibid., pp. 66-72.
46 Reading Weininger, or Edwards' description of Wittgenstein's ethical vision, one can't help but be reminded of Raskolnikov in Dostoyevsky's *Crime and Punishment.*
47 Letters to Ludwig von Ficker, in Luckhardt, *Wittgenstein, op. cit.*, n. 6, p. 91.
48 Edwards, *Ethics without Philosophy, op. cit.*, n. 7, p. 41.
49 Ibid., p. 70.
50 Phil., 2:5-11, with 'emptied himself' in place of the NIV's 'made himself nothing', and 'slave' in place of the NIV's 'servant'. Contrast this 'self-emptying', 'taking the very nature of a slave', with Weininger's lonely Kantian, far above all wanting-to-be-slave, being-able-to-be-slave, having-to-be-slave. Weininger's Kantian, far from emptying himself, is full of himself.
51 This transformation of our willing and acting is itself not an action that we can perform. Paul's talk of Christ's "humbling himself" is paradoxical insofar as it seems to suggest something that we can "imitate" by *doing something*, "humbling ourselves." But the transformation that is required has to be worked *upon* us. That this has occurred will be made manifest in the different character of those actions that we *do* perform.
52 Angelus Silesius, *Pélerin Chérubinique (Cherubinischer Wandersmann)*, ed. and trans. H. Plard (Paris: Aubier, 1946). This edition has French and German facing texts. I could not find a reliable and complete translation in English. I am responsible for all translations into English in this paper. Since my French is better than my German I have been guided by Plard's translation in reading the *Cherubinic Wanderer*, but I have in all cases translated the German and not the French translation.

The regard with which Wittgenstein held the *Cherubinic Wanderer* can be gleaned from the company it keeps in the following entry in his diary for 22 January 1931: 'Traffic with authors like Hamann, Kierkegaard, makes their editors presumptuous. This temptation would never be felt by the editors of the *Cherubinic Wanderer*, nor yet the *Confessions* of Augustine or a writing of Luther's': Ludwig Wittgenstein, *Denkbewegungen: Tagebücher 1930-1932, 1936-1937. Teil 1: Normalisierte Fassung*, ed. I. Somavilla (Innsbruck: Haymon Verlag, 1997), p. 41 (my translation).
53 Angelus Silesius, *Pélerin Chérubinique, op. cit.*, n. 52, p. 64.
54 Ibid, p. 68.
55 Ibid., pp. 346, 348.
56 Ibid., p. 104.
57 Ibid., p. 82.
58 Ibid., p. 130.
59 William James, *The Varieties of Religious Experience: a Study in Human Nature* (New York: Collier Books, 1961), p. 328.

60 Angelus Silesius, *Pélerin Chérubinique*, *op. cit.*, n. 52, p. 132.
61 Weil, *Gravity and Grace*, *op. cit.*, n. 5, p. 23.
62 Once again, this is not an action, but a transformation that is worked upon us. It is in a sense forced upon us by the very "logic" of solipsism, though it is always open to us to refuse to follow this through "strictly" to its conclusion, as we are drawn to do by the desire to maintain our illusions.
63 Paul Engelmann (ed.), *Letters from Ludwig Wittgenstein: With a Memoir* (New York: Horizon Press, 1968), p. 79.
64 John of the Cross, *Complete Works*, *op. cit.*, n. 32, p. 257.
65 Eph. 4: 22-24.

Chapter 4

Frege at Therapy*

Kelly Dean Jolley

> One thing I learned from Wittgenstein, in part from the *Tractatus*, but still more from personal contact, is that philosophical mistakes are often not refutable falsehoods but confusions; similarly the contrary insight cannot be conveyed in proper propositions with a truth-value. I offer as [an] instance of such insights Frege's distinction between concept and object…Such insights cannot be demonstrated as theses, but only conveyed dialectically; the dialectic process largely consists in the art…of reducing to patent nonsense the buried nonsense that is found in attempts to reject these insights. We cannot refute nonsense by a straight-forward logical process; as Frege said, logic cannot deal with nonsense, but only characterise it as being nonsense.[1]

> A simile that has been absorbed into the forms of our language produces a false appearance, and this disquiets us. 'But *this* isn't how it is!' – we say. 'Yet *this* is how is has to *be*!'[2]

Introduction

> There is not *a* philosophical method, though there are indeed methods, like different therapies.[3]

When struggling to understand – or just to cope with – Wittgenstein's talk of therapy, it is easy to feel abandoned at a spot in his thinking that is as far from Frege as any.

I propose to show that therapy *begins* in Frege. I first enter into the record two discussions of therapy by Rush Rhees – the two best discussions I know. I will not argue explicitly for these discussions as the best discussions of therapy. I will have to wait for another occasion to do that. However, the fact that what Rhees says so compellingly illuminates what Frege does is itself an implicit argument in favour of Rhees' discussion. I next examine Frege's response to Benno Kerry in 'On Concept and Object', demonstrating that the response lines up neatly with Rhees' discussions by recasting Frege's response in the terms provided by Rhees. I then consider the nature of the logical point Frege is making, and why therapy is internal to making the point. I end the paper with a Coda that examines the Tractarian predecessor-notion to therapy, elucidation. My tone will change somewhat in the Coda, since there my aim will be to show how to weave elucidation into the fabric of my earlier discussion.

Before I begin, I judge that a word about the way in which I have constructed the sections of the essay (preceding the Coda) is in order. Each section after the next begins, like this one, with remarks from Wittgenstein's later writings – in particular, from *Philosophical Investigations*. The remarks are not ornaments. In each section the remarks show Wittgenstein's indebtedness to Frege's thinking (and sometimes even to Frege's specific wordings) and show how Wittgenstein both enlarges and limbers Frege's thinking. Each section is written so that the remarks can show these things, at least upon reflection. Also, the sections build dialectically on one another, deepening and clarifying the structure of Frege's response to Kerry.

Finally, before I begin, I note that my examination of Frege's response to Kerry gets its light from my desire to understand Wittgenstein's talk of therapy. I do not claim that my examination of the response would harmonize with Frege's self-understanding of the response (although I also do not claim that it would not).[4] I want to highlight the lessons Wittgenstein learned from Frege – regardless of whether Frege self-consciously taught those lessons.

Rhees on Therapy

Here are Rhees' discussions of therapy:

> Philosophy as therapy: as though the philosopher's interest were in the personal disabilities of the perplexed: and as though he were not perplexed himself – as though philosophy were not discussion. Some remarks which Wittgenstein himself made are partly responsible for this. But he was suggesting an analogy with therapy; and he was doing this in order to bring out certain features in the method of philosophy: to show the difference between what you have to do here and what you would do in solving a problem in mathematics or in science. It was not a suggestion about what philosophy is interested in. If Wittgenstein spoke of 'treatment', it is the problem, or the question, that is treated – not the person raising it. It is not the personal malaise of the 'patient' which makes the perplexity or question important. What has led me to this perplexity is not my personal stupidity. Rather it is a tendency in language which could lead *anyone* there, and keeps leading people there. This is why Wittgenstein or anyone doing philosophy can understand the difficulty and the discussion of it. In this respect it is no more personal than the problems of science are. If it were a silly question – then I suppose it *would* be personal. What makes the questions deep and important is just that they are not this. And for this reason you do learn something from the discussion: it is not as though you were simply being restored to a normal state of mind. What makes you ask questions in philosophy is not a personal misfortune. And what can 'cure' you is philosophy, is discussion, is understanding. If you do not understand, then you will not have benefited in any way. And this shows that the benefit cannot rightly be described as a 'cure'.
>
> Wittgenstein speaks of 'temptations' we feel in reflecting on language, and especially on logic. (Perhaps we keep wondering whether the grammar and the concepts we use allow us to grasp things as they really are, for instance.) But he is not saying something about human nature, or the special weaknesses of those who do philosophy. He sometimes spoke of analogies between philosophy and psychoanalysis, but these are analogies in *method*. The functional disorders which philosophy treats appear as delusions and dreams of our language....And what he says about 'tackling' philosophical difficulties might suggest, again and again, some analogy with treatment. To root out the

> difficulty we have to start thinking about these things in a new way; you have to call to mind something you have been doing all along – something which (for special reasons) it is *hard* to call to mind; you have to bring together practices which you know and take for granted, so that each of them appears in a different light; and this cannot be done all at once: we are concerned with problems whose answer you cannot reach just by thinking, but through practice....[5]

I will make use of these discussions below.

Kerry's Folly

> But the diviner would say: 'Surely you know what ['I feel in my hand that the water is three feet under the ground'] means. You know what 'three feet under the ground' means, and you know what 'I feel' means!' but I should answer him: I know what a word means *in certain contexts*. Thus I understand the phrase, 'three feet under the ground', say, in the connections 'The measurement has shown that the water runs three feet under the ground'....[6]

Kerry, commenting on Frege's *The Foundations of Arithmetic*, objects to Frege's distinction between concept and object. In Kerry's words, Frege argues for 'the view that the properties of being a concept and being an object are mutually exclusive'.[7] Kerry says that such a view is no more plausible than is the claim that the relation of father to son 'were one that could not be further reduced, that a man could not be at once a father and a son (though of course not, e.g., father of the man whose son he was).'[8] Kerry reckons that he can rather simply show that Frege is mistaken. He offers the following example: 'The concept "horse" is a concept easily attained'. Kerry claims that in this sentence *a concept* – namely, the concept 'horse' – is playing the role of an *object* (just as a father might play the role of a son). Thus, the sentence shows that Frege's view is mistaken.

The opening paragraphs of Frege's 'On Concept and Object' are crucial to understanding Frege's response to Kerry. Frege accuses Kerry of confusing the psychological use of 'concept' with its logical use. Anyone familiar with the introduction to The *Foundations of Arithmetic* will immediately recognize that Frege is accusing Kerry of failing to keep to one of Frege's three fundamental principles: *always to separate sharply the psychological from the logical, the subjective from the objective*. Frege's comment on this principle in his *Foundations* is that, in compliance with it, he has 'used the word "idea" always in a psychological sense, and [has] distinguished ideas from concepts and from objects'.[9] In 'On Concept and Object', Frege explains Kerry's misunderstanding as follows:

> The word 'concept' is used in various ways; its sense is sometimes psychological, sometimes logical, and sometimes perhaps a confused mixture of both. Since this license exists, it is natural to restrict it by requiring that when once a usage is adopted it shall be maintained. What I decided was to keep strictly to a logical use....Agreement about the mode of expression will easily be reached when once it is recognized that there is something that deserves a special term.

> It seems to me that Kerry's misunderstanding results from his unintentionally confusing his own usage of the word 'concept' with mine. This readily gives rise to contradictions, for which my usage is not to blame.[10]

Frege sees Kerry as confused about the use of 'concept'. Confused, Kerry generates contradictions and then blames the contradictions on Frege's use of the word.

Because he is confused about the use of 'concept', Kerry is confused by Frege's distinction between concept and object. Frege claims that the distinction is absolute; Kerry denies it. Here Kerry transgresses against another fundamental principle of Frege's: *never to lose sight of the distinction between concept and object*. (Kerry would of course say that he has not lost sight of the distinction but only that he is denying that the distinction is absolute; Frege would respond by saying that to deny that the distinction is absolute is to lose sight of the distinction. The reasons for Frege's response will become clearer below.)

The importance of recognizing that Kerry has failed to keep to these two principles is that it prompts the suspicion that Kerry has also failed to keep to the third of Frege's principles, the so-called Context Principle: *never to ask for the meaning of a word in isolation, but only in the context of a proposition.*[11] When we reach the heart of Frege's response to Kerry, this suspicion is confirmed. To see that it is, reconsider Kerry's counterexample to Frege's claim that the concept/object distinction is absolute:

> (K) The concept 'horse' is a concept easily attained.

(K) is supposed to be a counterexample, since 'the concept "horse"' appears in it in the place of an object-expression. Given the way in which Kerry misunderstands Frege, Kerry takes (K) to show that the concept-expression 'the concept "horse"' is playing the role of an object-expression; hence, Kerry takes (K) to show that the concept/object distinction is not absolute.

That Kerry misunderstands Frege is easy to see after we remind ourselves how Frege understands the segmenting of statements like:

> (F) Silver is a horse.

For Frege, the sentence segments into an object-expression – 'Silver' – and a concept-expression – 'is a horse'. Frege thinks of object-expressions as *saturated* and of concept expressions as *unsaturated*. Frege represents the unsaturatedness of the concept-expression by writing it with a blank spot in it: '( ) is a horse'.[12] The blank spot is a notational reminder of the difference in logical role between the concept-expression and the object-expression. When the object-expression saturates the unsaturated concept-expression, we get the statement (F). The problem Kerry faces is that of showing that an unsaturated concept-expression can substitute for a saturated object-expression: that is, Kerry needs to be able to substitute *salva congruitate* an expression with a blank spot for one without. But this sort of substitution fails. Substituting an expression with a blank spot for one without it turns a statement into a mere farrago of words.[13]

Why does Kerry fail to see this? Because he transgresses the Context Principle. Kerry thinks about the words 'the concept "horse"' outside the context of a proposition and thinks that they must be a concept-expression. They must be because, to him, the words seem to assign that role to themselves. With their logical role outside the context of a proposition assigned, Kerry finds a propositional context for the words, namely (K). Kerry segments (K) as Frege would, treating the words 'the concept "horse"' as an object-expression. Kerry then thinks he has counterexampled Frege since he (Kerry) thinks he has substituted words that play the role of a concept-expression into the role of an object-expression: Kerry understands the words 'the concept "horse"' to play the role of a concept-expression outside the context of (K) and, while retaining that role, to also play the role of an object-expression in (K). If Kerry understands correctly, (K) would indeed be a counterexample.

It should be clear now why Frege thinks Kerry misunderstands, and why (K) fails to worry Frege. For Frege, the logical role played by the words 'the concept "horse"' is something that cannot be decided by considering the words divorced from context, since only in contextual wedlock do words play a logical role and only in contextual wedlock are words object-expressions or concept-expressions. And, in contextual wedlock, words will be absolutely one or the other, and not both.

Kerry wants to be able both to decide the role of words outside the context of a proposition and then, holding that role constant, to put the words into a different role in the context of a proposition. This shows, I think, why Frege mentioned the possibility of a confused mixture of psychological and logical senses of 'concept'. For Frege, the words 'the concept "horse"' in (K) are used as an object-expression, so, given what (K) says, the word 'concept' has a psychological sense (say, it refers to an object in someone's mind).[14] Frege has no objection to the word having such a sense, so long as we do not confusedly mix this sense with his own sense of 'concept'. But Kerry does confusedly mix them: he ignores the Context Principle in attempting to assign a role to the words outside the context of a proposition, but he also heeds the Principle in attempting to assign a role to the words in the context of a proposition. Kerry thinks that the words 'the concept "horse"' can have a Fregean, logical sense outside the context of (K) and then maintain that sense while having also a psychological sense inside the context of (K).

Delusions of the Language

> A *picture* held us captive. And we could not get outside it, for it lay in our language and language seemed to repeat it to us inexorably.[15]

I want now to make use of Rhees' discussions of therapy, as they apply to Frege's response to Kerry. The first thing to notice is that Kerry's plight is the result of what Rhees calls a 'delusion or dream' of the language. Kerry confronts the words 'the concept "horse"', and is deluded by the words themselves: he allows those words to suggest that they can assign themselves a logical role, that they can decide whether they are a concept-expression or an object-expression, outside the context of a proposition.

Frege is himself alert to just such delusions of the language. In 'Comments on *Sinn* and *Bedeutung*', he writes:

> Such being the essence of a concept, there is now a great difficulty in the way of expressing ourselves correctly and making ourselves understood. If I want to speak of a concept, language, with an almost irresistable force, compels me to use an inappropriate expression which obscures – I might almost say falsifies – the thought. One would assume, on the basis of its analogy with other expressions, that if I say 'the concept "equilateral triangle"' I am designating a concept, just as I am of course naming a planet if I say 'the planet Neptune'. But this is not the case; for we do not have anything with… [an unsaturated] nature.[16]

The similarity of the compulsion Frege describes and the one to which Kerry succumbs is clear. Notice first that Frege blames language for the compulsion '…language, with an almost irresistible force, compels me…'. It is in casting blame on language that Frege explains what I have taken Rhees to mean when talking of 'delusions' of the language: 'One would assume…that if I say "the concept 'equilateral triangle"' I am designating a concept'. Here we might imagine Kerry chiming in 'Yes! After all, the words are the words "the *concept* 'horse"'! They must designate a concept!', 'But', as Frege would say, 'this is not the case; for we do not have anything with…[an unsaturated] nature.' That is, the words cannot assign themselves the logical role of a concept-expression. Given the way in which such words are used, as in (K), the words have the role of an object-expression, despite the fact that the word 'concept' is one of their fellow travelers.[17] The delusional power of the language lies in the analogies it suggests to its users. As Frege mentions, seeing in the words 'the concept "equilateral triangle"' an analogy to the words 'the planet Neptune' is almost irresistible. Our familiarity with uses of the words 'the planet Neptune' in such sentences as

> (F') The planet Neptune is distant from Earth.

makes properly understanding the logical role of (shifting examples) 'the concept "horse"' difficult. We want to see the words as designating a concept, analogous to the way in which the words 'the planet Neptune' designate a planet in (F'). But, since we think that the words designate a concept, we think that they must play the role of a concept-expression. (However, insofar as the words designate a concept, they play the role of an object-expression – as do the words 'the planet "Neptune"' in (F').) If we think of the words as assigning themselves a logical role, then as Kerry did and as Frege confesses being almost compelled to do, we succumb to a delusion of the language.

Notice next that Frege is describing a delusion that is anything but personal: Kerry is preyed upon by it, but Frege almost is, too. The tendency in the language that preys upon Kerry could prey upon anyone. Kerry's problem is not a personal shortcoming of his own; he is not displaying a special weakness. The analogy that deludes him is there – in the language – awaiting anyone who is unwary.

The Nature of Logical Investigations

> Our investigation is therefore a grammatical one. Such an investigation sheds light on our problem by clearing misunderstandings away. Misunderstandings concerning the use of words, caused, among other things, by certain analogies between forms of expression in different regions of language. – Some of them can be removed by substituting one form of expression for another; this may be called an 'analysis' of our forms of expression, for the process is sometimes like one of taking a thing apart.
>
> These are, of course, not empirical problems; they are solved, rather, by looking in to the workings of our language, and that in such a way as to make us recognize those workings: *in despite of* an urge to misunderstand them. The problems are solved, not by giving new information, but by arranging what we have always known. Philosophy is a battle against the bewitchment of our intelligence by means of language.[18]

Kerry's problem, the delusion of language to which he succumbs, is no more his personal problem than is a problem in science; in this way, the delusion and a problem in science are alike. But, in other ways, the delusion is significantly unlike a problem in science. Kerry's problem is a logical problem, not an empirical problem. Frege's response to Kerry's problem is a logical response – call the response a logical investigation. What Kerry misunderstands is the use of the words 'the concept "horse"'. Frege responds by reminding Kerry how the words are used in (K) and how they fail to be used outside of (K). Such reminding will not solve an empirical problem: being reminded in this way of how words are used will not solve problems in, say, physics or chemistry.[19]

What Frege reminds Kerry of is something that Kerry has been doing all along. Kerry has been using words like 'the concept "horse"' as object-expressions all his life. He has observed the absolute distinction between such expressions and concept-expressions. That he has done this is revealed by the presumably largely correct inferences he has drawn over the course of his life. But Kerry finds it hard to call to mind what he has been doing for so long. Under the pressure of reflection, Kerry cannot bring himself to keep to the Context Principle, cannot bring himself to refuse to ask for the meaning of certain words in isolation from propositions. Confronted by the words 'the concept "horse"', Kerry thinks he has discovered the logical role of the words by simply thinking about the words themselves. To prevent this, Frege (so to speak) discusses with Kerry. He urges Kerry to think about the words and their logical role in a different way: He asks Kerry to consider the logical role the words play in (K), and to realize that nothing gained from reading the words in isolation from (K) (or some other proposition) can reveal their logical role. So, in order to help Kerry solve his problem, Frege needs to get Kerry to do something – to actually segment (K) – and not to just think about the words that appear in (K), and especially not to just think about them in isolation from (K). Learning to segment sentences, learning to keep to the Context Principle, is something that takes practice. The analogies in language that tempt us to transgress against the Principle crowd us from every side. The problems that prey upon us as a result of the analogies are not problems we solve by just thinking; we solve them by building up our resistance to being compelled by the analogies, by continually reminding ourselves that words cannot assign themselves a logical role. Put another way, we solve these problems by forcing ourselves not only continually to remember, but continually to keep to the

Context Principle. Frege's response works to make Kerry aware of how such problems are solved, to show Kerry how to free himself from 'the concept "horse"' fly-bottle. Freeing himself will require Kerry to understand how he became entrapped, as O.K. Bousma makes clear:

> That fly that was let out of the fly-bottle understands how he got in there, since the condition of his being let out is that he should understand that. And now he can fly in and out as he likes. It is no longer a fly-bottle for him. He can now buzz in and out enjoying the structure of the bottle.[20]

Making us aware of how such problems arise and are solved, getting us to understand how we entrap ourselves and free ourselves from the fly-bottle, is the burden of Frege's response. It is the burden of the therapist. As Frege writes in 'Sources of Knowledge of Mathematics and the Mathematical Natural Sciences':

> It is extremely difficult, perhaps impossible, to test every expression offered us by language to see whether it is logically innocuous. So a great part of the work of a philosopher consists – or at least ought to consist – in a struggle against language. But perhaps only a few people are aware of the need for this…The difficulties which…language entangles us in are incalculable.[21]

So therapy is internal to logical investigation. Therapy is internal to logical investigation because the scenes of logical investigation are the scenes of trespass against the Context Principle – the scenes of entanglement in language. Our entanglements in language do not result in falsehood but rather in nonsense. And therapy is responsive to nonsense – uniquely responsive to nonsense and responsive uniquely to nonsense. Therapy inoculates us against language.

Summary

> The results of philosophy are the uncovering of one or another piece of plain nonsense and of bumps that the understanding has got by running its head up against the limits of language. These bumps make us see the value of the discovery.[22]

It is all too easy to think that Frege's response to Kerry is peculiarly unresponsive. It is all too easy to think that deciding for or against Frege or Kerry is a matter of deciding whether or not (K) is, or can be, meaningful. But that decision always comes one moment too late in the discussion. The decision needed is not whether (K) is or can be meaningful, but rather whether Kerry can mean by (K) what he needs to mean by it, given his intention to counterexample Frege.[23] To rush past that decision in order to decide whether (K) is, or can be, meaningful is to rush past the Context Principle – to rush past the need for and the possibility of therapy. It is to be guilty of Kerry's Folly.

Therapy can be found in Frege. Wittgenstein found it there.

Coda: Tractarian Therapy

> *Tractatus Logico-Philosophicus* 4.112 The object of philosophy is the logical clarification of thoughts. Philosophy is not a theory but an activity. A philosophical work consists essentially of elucidations. The result of philosophy is not a number of 'philosophical propositions', but to make propositions clear. Philosophy should make clear and delimit sharply the thoughts which are, as it were, opaque and blurred.
> 4.1212 What *can* be shown *cannot* be said.
> 4.1213 Now we understand our feeling that we are in possession of the right logical conception, if only all is right in our symbolism.[24]

It may seem that I am guilty of a peculiar philosophico-literay fallacy; call it 'the fallacy of extruded middle'. After all, I've moved from Frege's 'Concept and Object' to Wittgenstein's *Philosophical Investigations* and said nothing about Wittgenstein's intervening work, particularly his work in the *Tractatus*. One reason for so far skipping the *Tractatus* is the fact that Frege's influence on it is more widely recognized than is his influence on *Philosophical Investigations*. But, having made my case for Frege's influence on *Philosophical Investigations*, I've reached a natural point from which to say a little about Frege's influence on the *Tractatus* – particularly on the *Tractatus*' counterpart to therapy, *elucidation*. It is also, in many ways, a better point: *Philosophical Investigations* sheds at least as much light back on the *Tractatus* as the *Tractatus* does forward on *Philosophical Investigations*. (Wittgenstein was right to want the books bound together.[25])

To understand elucidation, we must understand two distinctions in the *Tractatus*: the distinction between sign and symbol, and the distinction between proper concepts and formal concepts. I will briefly explain these distinctions, and then show how they enter into Frege's argument with Kerry.

Wittgenstein notes that in order to recognize the symbol in the sign, we must 'consider the significant use' of the sign (TLP 3.326). In a passage from the 'Notes Dictated to Moore', Wittgenstein says that '…in "aRb", "R" is *not* a symbol, but *that* "R" is between one name and another symbolises.'[26] What is crucial in this passage is the way in which Wittgenstein uses logical syntactic employment ('logical syntactic employment' is Wittgenstein's 3.327 gloss on 3.326's 'significant use') to discern the symbol in the sign 'R'. The sign 'R' on its own does not disclose a symbol; it does not symbolize. Rather, 'R' standing between two names symbolizes. This connects up intimately with Frege's Context Principle. Wittgenstein is echoing the Context Principle, but in Tractarian terms: 'Never ask what symbol is in a sign in isolation, but only in the context of the sign's logical syntactic application.' (Thus is the Context Principle inceptive of the sign/symbol distinction.) The sign 'R' symbolizes in 'aRb', but the sign 'R' in isolation does not (yet) symbolize. In isolation from a proposition, a sign is divorced from contextual wedlock; it is divorced from the logical syntactic employment, in virtue of which it symbolizes.[27]

At TLP 4.126 Wittgenstein explains why he introduces the expression 'formal concept': 'I introduce this expression in order to make clear the confusion of formal concepts with proper concepts which runs through the whole of the old logic.' Wittgenstein goes on to say (TLP 4.127) that a propositional variable 'signifies the formal concept, and its values signify the objects which fall under this concept'.[28] For

Wittgenstein, both Frege's *objects* and *concepts* are formal concepts – in other words, each is a way of talking about the values of particular propositional variables.[29] This has the consequence that neither *objects* nor *concepts* can be talked about: 'That anything falls under a formal concept as an object belonging to it, cannot be expressed by a proposition....But it is shown in the symbol for the object itself' (TLP 4.126). In other words, whether something is an *object* or a *concept* is decided not by claims about it, but by recognizing the propositional variable of which it is the value.[30]

When Kerry treats (K) as a counterexample to Frege, Kerry confuses himself (putting this in the Tractarian terms that I've been explaining) about both signs and symbols, and proper concepts and formal concepts. Kerry thinks that a sign can determine its own symbol – that is, he thinks that the words 'the concept "horse"' provide themselves with the logical syntactical role of a *concept*. Kerry thinks that the words 'the concept "horse"', on their own and outside of contextual wedlock, determine their logical form – determine the propositional variable of which they are the value. (This shows how the two distinctions – sign/symbol and proper concept/formal concept – interrelate: to see the symbol in the sign is to recognize the propositional variable of which the sign is the value.) Kerry thinks that the words 'the concept "horse"', outside the context of a proposition, assign themselves the value of a *concept* propositional variable. Again, Kerry is clear enough about the Context Principle to realize that the words 'the concept "horse"' are, inside the context of (K), the value of an *object* propositional variable. Kerry needs this to work in order to show that the distinction between *concept* and *object* is not absolute. And so Kerry manages confusedly to mix the logical and psychological, as was shown before. (As the value of an *object* propositional variable, the words 'the concept "horse"' have a psychological sense.)[31]

Frege's response to Kerry is the model of, and for, Tractarian *elucidation* – the logical clarification of thoughts (TLP 4.112). Frege tries to get Kerry to understand that the symbol in the sign cannot be seen outside the context of significant use, outside of logical syntactic employment. He tries to get Kerry to recognize that nothing gained from staring at 'the concept "horse"' in isolation from some logical syntactic employment can reveal of what propositional variable the words are the value. In essence, what Frege asks Kerry to see – and which Kerry does see, in a way – is that 'Every variable can be conceived as a propositional variable' (TLP 3.314). Kerry sees that to fall under a formal concept is to be the value of a propositional variable, *and* that to be the value of a propositional variable is to play a particular logical syntactic role in a proposition. Getting Kerry clearly to see this is to get him to stop theorizing about the words of (K) and to do something – actually to segment (K). Kerry finds it hard to get completely clear about the logical form of (K) because he thinks of logical clarification – elucidation – as a theory and not as an activity: he thinks that elucidating culminates in a series of philosophical propositions that assign (formal) properties to propositions or parts of propositions (cf. TLP 4.124 and 4.126). Instead, as Frege's response shows, elucidating (K) is a matter of segmenting (K), of making (K) clear, of sharply delimiting the thought that (K) expresses. What Frege tries to show Kerry – and did show Wittgenstein – is that elucidating (K) is not a matter of having some number of thoughts about (K)'s logical form, but is rather a matter of thinking clearly when thinking (K). It is not a matter of thinking the form of some particular thought, but of thinking a thought with some particular form – activity, not theory.

How does Wittgenstein get from Frege's response to Kerry to elucidation and then to therapy? Here's a passage from Anscombe that can serve as both the beginning of the answer to that question and as a frame for some of the details I have supplied concerning the *Tractatus*:

> ...Frege came to think that if something is a concept, we cannot correctly say that it is a concept – i.e., predicate the term 'concept' of it – because an expression for a concept can significantly occur only in the place of a predicate, not as a subject of the predicate 'concept'. This doctrine was what Wittgenstein expressed by saying: 'That anything falls under a formal concept as an object belonging to it, cannot be expressed by a proposition. But it is shown in the symbol for the object itself. (4.126)'; e.g., if something falls under the 'formal' concept *concept* or *property*, this is shown by the predicative character of the sign we use for that 'something'; and again, a variable relating to properties will be one that we take as having one or more argument-places.
>
> In Wittgenstein...the notion of a 'formal' concept, a concept that cannot be properly expressed by a predicate or general term, but only by the way we apply the corresponding sort of sign, is extended much more widely than this. Not only 'concept', 'function', 'object', but also 'number', 'fact', 'complex', are formal concepts.[32]

Anscombe describes nicely the way in which Wittgenstein, in the *Tractatus*, inherits and expands the possibilities of elucidation first displayed in Frege's response to Kerry. By the time Wittgenstein writes *Philosophical Investigations* he will expand the possibilities yet again (quoting Michael Thompson):

> It is one of the lessons taught by Wittgenstein, if I understand him, that we must recognize many intuitively more determinate distinctions *of the sort Frege introduced* [with his distinction between *concept* and *object*]. Wittgenstein of course calls the corresponding sort of distinction...a 'grammatical difference'.[33]

Wittgenstein moves to elucidation and from there to therapy by reflecting on Frege's response to Kerry, by enlarging and limbering that response.

Anscombe claims that '...almost all that has been published about [the *Tractatus*] has been wildly irrelevant. If this has had any one cause, that cause has been the neglect of Frege and the new direction he gave to philosophy.'[34] Anscombe would be right to claim this of *Philosophical Investigations*, too – particularly about therapy.

Notes

* I thank Michael Watkins, Jody Graham, Pierluigi Miraglia, Elizabeth Brake, Gary Cesarz, Alice Crary, Eric Marcus and especially Ram Neta for helpful comments and discussions. This paper was delivered as the Presidential Address at the 2000 Alabama Philosophical Society, and I thank also the members of that audience.

1 Peter Geach, *Philosophical Encounters*, ed., H. Lewis (Dordrecht: Kluwer, 1991), pp. 13–14.

2 Ludwig Wittgenstein, *Philosophical Investigations*, eds, R. Rhees and G.E.M. Anscombe, trans. G.E.M. Anscombe, 3rd edn (Oxford: Blackwell, 1958), § 112.

3 Ibid., § 133.
4 In another discussion, I would argue that Frege's self-understanding is conflicted: he sometimes understands his response to Kerry (and his handling of related matters) as Wittgenstein does; he sometimes understands it in a way inconsistent with the way in which Wittgenstein does. This conflict manifests itself throughout Frege's writings.
5 R. Rhees, 'A Symposium: Assessments of the Man and the Philosopher', in *Ludwig Wittgenstein*, ed., K.T. Fann (New York: Delta Publishing, 1967), pp. 77–78.
6 Ludwig Wittgenstein, *The Blue and Brown Books*, 2nd edn (Oxford: Basil Blackwell, 1969), pp. 9–10.
7 Gottlob Frege, 'On Concept and Object', in *Translations from the Philosophical Writings of Gottlob Frege*, ed. Peter Geach and Max Black (Oxford: Blackwell, 1977), p. 43.
8 Frege, *Translations from the Philosophical Writings of Gottlob Frege*, *op. cit.*, p. 43.
9 Gottlob Frege, *The Foundations of Arithmetic* (Oxford: Blackwell, 1968), p. x.
10 Frege, *Translations from the Philosophical Writings of Gottlob Frege*, *op. cit.*, p. 43.
11 Kerry's failure at least to address these principles is made more serious by the fact that he wrote in response to the *Foundations*.
12 Keep in mind that speaking of '( ) is a horse' as a concept-expression only makes sense where we are abstracting from a particular statement(s), like (F); the same is true of speaking of 'Silver' as an object-expression.
13 '...[O]bjects and concepts are fundamentally different and cannot stand in for one another. And the same goes for the corresponding words or signs. Proper names cannot really be used as predicates. Where they might seem to be, we find on looking more closely that the sense is such that they only form part of the predicate: concepts cannot stand in the same relations as objects. It would not be false, but impossible to think of them as doing so': Frege, 'Comments on *Sinn* and *Bedeutung*', in *The Frege Reader*, ed., M. Beaney (Oxford: Blackwell, 1997), pp. 174–75. My understanding of Kerry's problem here has been influenced by the final chapter of Joan Weiner's *Frege in Perspective* (Ithaca, NY: Cornell University Press, 1991). I again warn the reader that, with this description of the problem Kerry faces, I mount a dialectical ladder that I will not kick away until just before the Coda. Only when I have progressed that far will it be clear what my description here means and fails to mean. (The same mounting and kicking-away will be repeated in the Coda, too.) As shall become clear, there is no coherent description of the problem Kerry faces; and *that* is ultimately the problem he faces.
14 It is important to remember that I am talking about the words *in (K)*. The words, 'the concept "horse"', used as an object-expression in some other proposition, need not have a psychological sense. But Kerry intends them to have such a sense in (K).
15 Wittgenstein, *Philosophical Investigations*, *op. cit.*, n. 2, § 115.
16 Frege, *The Frege Reader*, *op. cit.*, n. 13, p. 174.
17 Someone might here object that there is no delusion in the language, but rather that Frege was deluded in his choice of terminology: he should never have used the term 'concept'. But this objection – a version of which Frege responds to in the closing paragraphs of 'On Concept and Object' – fails to appreciate that any term Frege might have chosen would create the same sort of confusion, since an analog of (K) could be just as well be produced for it. As Frege might have said, 'The difficulty can indeed be shifted, but not avoided'.
18 Wittgenstein, *Philosophical Investigations*, *op. cit.*, n. 2, §§ 90, 109.
19 A question worth asking: will reminders of how words are used solve problems in *empirical linguistics*? The briefest answer worth giving is 'No – use is not *usage*'. What Frege is reminding Kerry of is not a bit of empirical linguistics; he is not reminding Kerry of the usage of certain German words. If Frege were doing so, he would be doing something much more like what Kerry takes him to be doing.
20 O.K. Bouwsma, 'The Blue Book', in *Philosophical Essays* (Lincoln: University of Nebraska Press, 1965), p. 186.

21 Frege, *The Frege Reader*, *op. cit.*, n. 13, pp. 369–70.
22 Wittgenstein, *Philosophical Investigations*, *op. cit.*, n. 2, § 119.
23 Putting the point in terms borrowed from Stanley Cavell, Kerry's problem is that not that (K) means something incoherent, but rather that Kerry means (K) incoherently – Kerry has been 'prompted to insistent emptiness'. See S. Cavell, *The Claim of Reason: Wittgenstein, Skepticism, Morality and Tragedy* (Oxford: Oxford University Press, 1979), p. 336.
24 All in-text references to Wittgenstein's *Tractatus Logico-Philosophicus* are hereafter denoted as 'TLP'.
25 Wittgenstein says that his later way of thinking can be seen in the right light 'only by contrast with and against the background of [his] old way of thinking': Ludwig Wittgenstein, *Philosophical Investigations*, *op. cit.*, p. x. The reverse is also true, I think.
26 Ludwig Wittgenstein, *Notebooks, 1914–1916*, eds, G.H. von Wright and G.E.M. Anscombe, trans. G.E.M. Anscombe (New York: Harper and Row, 1969), p. 108.
27 That the Context Principle is inceptive of the sign/symbol distinction reveals something important about the Context Principle generally and about its role in the *Tractatus* specifically: the Context Principle is the methodological counterpart of the unity of the proposition. (Cf. TLP 3.14–3.144.)
28 I should point out that both my reader and the reader of the *Tractatus* must be cautious of Wittgenstein's use of 'object' in TLP 4.126ff. This use is peculiar and potentially misleading. It may help the reader to compare this use of 'object' with that in *Philosophical Investigations* § 373: 'Grammar tells what kind of object anything is.' For more help, see G.E.M. Anscombe's *An Introduction to Wittgenstein's Tractatus* (London: Hutchinson, 1959), p. 122, n. 1.
29 Although I have not done so until now, I am adopting the typographical expedient of italicizing 'object' and 'concept' when using the terms for formal concepts. The expedient is meant to help with the potentially confusing talk of 'object' in TLP 4.126ff (see n. 28).
30 Appreciating the philosophico-logical achievement of the *Tractatus* is a matter of seeing how Wittgenstein's claim that 'the object of philosophy is the logical clarification of thoughts', and so on, (TLP 4.112) fits together with two other claims: 'The "logical constants" do not represent' (TLP 4.0312) and 'Every variable is the sign of a formal concept' (TLP 4.1271).
31 It is in treating 'the concept "horse"' as assigning itself the value of a *concept* propositional variable while being assigned the value of an *object* propositional variable that Kerry renders (K) opaque and blurred – and that he renders it opaque and blurred in this way sheds light on how and why it is that philosophical confusions arise (TLP 3.323–3.325) and why it is that 'all the propositions of our colloquial language are actually, just as they are, completely in order' (TLP 5.5563).
32 Anscombe, *Introduction*, *op. cit.*, n. 28, pp. 122–23.
33 Michael Thompson, 'The Representation of Life' in *Virtues and Reasons*, eds, R. Hursthouse, G. Lawrence and W. Quinn (Oxford: Oxford University Press, 1995), p. 249, n. 2. The initial italics are Thompson's.
34 Anscombe, *Introduction*, *op. cit.*, n. 28, p. 12.

- The roots of the therapeutic approach are in Frege.
- Whether or not this conforms to F's self-underst. (which seems to have inner conflicts) W draws such lessons from F.
- R's remarks on W are used to clarify the therapy analogy.
- Phil. probs are not personal probs, but lang. leads us to them.
- Eluc. in TLP is a forerunner of therapy.

Chapter 5

'Making Sense' of Nonsense: Conant and Diamond Read Wittgenstein's *Tractatus*

Diarmuid Costello*

> I ought to be no more than a mirror, in which my reader can see his own thinking with all its deformities so that, helped in this way, he can put it right.[1]

Introduction: The Debate over TLP 6.54[2]

This paper focuses on the debate over Wittgenstein's *Tractatus*, specifically the question as to why Wittgenstein would have written a book which, by its own lights, has to be considered largely nonsensical. For not only is the *Tractatus* a book that closes by acknowledging that it is nonsense, it makes recognizing that it is, a requirement of 'understanding' it. Or rather, it makes such recognition a criterion of having understood its author:

> My propositions serve as elucidations in the following way: anyone who understands me eventually recognizes them as nonsensical, when he has used them – as steps – to climb up beyond them. (He must, so to speak, throw away the ladder after he has climbed up it.)
>
> He must transcend these propositions, and then he will see the world aright. (TLP 6.54)

Not surprisingly, this remark has occasioned much controversy amongst the book's readers. For how can nonsense be interpreted? What must be grasped in order to do so? Indeed, if it is nonsense, can there be anything *to* grasp? Until recently, interpreters tended to dilute the force of this paradox: the book's propositions may be nonsense, but they are 'illuminating' all the same. They are a special kind of nonsense that results when philosophers 'violate the logical syntax of language' in

dilute

a misguided attempt to 'say what can only be shown'. Following James Conant, I shall call this the orthodox account and I will take P.M.S. Hacker's *Insight and Illusion* as representative.[3]

The debate I focus on here began when Cora Diamond rejected the idea of 'deep' nonsense underpinning such accounts. Diamond argued that this is an incoherent notion that fails to do justice to the *Tractatus*.[4] When Wittgenstein urges his readers to recognize his propositions as nonsense, he wants us to see them for what they are – that is, plain, unvarnished nonsense. The test of this reading is whether it can make sense of the book without collapsing back into the positivist reception from which its orthodox commentators had sought to rescue it. In addition, it will owe us some explanation of what Wittgenstein might have hoped to *achieve* by knowingly writing a book of nonsense. More recently, Diamond's argument has been taken up by James Conant, who has added a perfectionist dimension to her reading. Between them, they have sought to contest all *Tractatus* scholarship to date. Moreover, by doing so, their work calls for a re-evaluation of what we might call Wittgenstein's 'philosophical authorship' as a whole, particularly the standard view of the *Philosophical Investigations* as a critique of the discredited 'doctrines' of the *Tractatus*. For only if Wittgenstein really was advancing a metaphysical theory about the relation between language and world in the *Tractatus* does such a view make any sense.[5] If not – and this is the perfectionist wager – the *Tractatus* may yet turn out to have more in common with the therapeutic aims of his later work than is generally realized. Hence two very different views of Wittgenstein's philosophy as a whole turn upon whether or not the Conant–Diamond reading of the *Tractatus* goes through. Whether it does is what I examine in this paper.

I begin by outlining the relevant aspects of the orthodox account, focusing on Hacker's exegesis of the distinction between saying and showing. I pay particular attention to Hacker's claim that this distinction gives rise to a special kind of nonsense that explains Wittgenstein's remarks in TLP 6.54. In the latter half of the paper, I consider the account put forward by Conant and Diamond. Against Hacker's view that the *Tractatus* attempts to communicate ineffable metaphysical truths about the relation between language (or thought) and the world, Conant and Diamond focus on the self-reflexive realization to which they believe Wittgenstein's book is intended to bring in its readers. Thus, unlike orthodox interpreters, Conant and Diamond argue that the *Tractatus* is motivated by Wittgenstein's abiding belief that philosophy consists in unwittingly taking nonsense for sense. On this view of the *Tractatus*, it is Wittgenstein's first attempt to bring his readers to existentially appropriate this insight; and it is the book's failure to do so that has the most fundamental implications for the method of his later philosophy.

The Background to the Debate: Sense, Senselessness and Nonsense in the *Tractatus*

Faced with Wittgenstein's consignment of his propositions to the realm of nonsense at TLP 6.54, it is necessary to clarify what Wittgenstein means by 'nonsense'. In the *Tractatus* Wittgenstein distinguishes three kinds of proposition: genuine propositions with a sense; senseless but nonetheless well-formed propositions; and nonsense. Genuine propositions are bipolar. They are either true or false depending on how things stand in the world. They picture a possible state of affairs. They do not require that the state of affairs depicted exist, only that it could exist. False propositions are false because they fail to depict how things stand in the world. Wittgenstein's picture theory of meaning entails that all propositions show what they say.[6] Their articulation represents the articulation of elements in the world. Wittgenstein calls this the pictorial form of the proposition. A proposition may not appear to be a picture of the state of affairs it represents, but that is because its apparent (grammatical) form need not be its real logical form and, generally, it isn't. Nonetheless, underpinning the theory is the notion of a hard and fast correlation between language and world. This may be disguised by grammatical form, but it is there all the same, and may be brought out by analysis.

Unlike meaningful propositions, senseless propositions are propositions constructed in such a way that they annul their own content – by either asserting *and* denying it simultaneously (*p*. ~*p*) or asserting that it either is *or* that isn't the case (*p* v ~*p*). As a result they fail to pick out a particular state of affairs from a range of possible states of affairs in the world. They say nothing. Unlike genuine propositions, their truth or falsity is not left open for the world to determine. They are unconditionally true or unconditionally false. Senseless propositions are either tautologies or contradictions: the former are necessarily true and the latter are necessarily false regardless of how things stand in the world. Nonetheless, the picture theory of meaning entails that they too *show what they say*. However, given that senseless propositions are constructed in such a way as to cancel out the meaning of their constituent parts, claiming that they show what they say amounts to claiming that they *show that they say nothing*. They exhibit their own internal structure, and it is in virtue of this structure that they say nothing.

Finally, the *Tractatus* holds that nonsense is all that lies on the far side of meaningful discourse. Nonsense is simply the failure to express a sense. Wittgenstein claims in the book's preface that his aim is to draw a limit to the expression of thoughts working outwards through what can be said. This is possible, according to the *Tractatus*, because the realms of what can be said and what can be thought are coextensive.[7] The point is not to *impose* a limit on what can be thought, but rather to find some means of locating it. Such a limit should not be understood as a limitation: it must encompass everything that can be said, excluding nothing.[8] If Wittgenstein's book succeeds in drawing such a limit it will have thereby demarcated what is neither thought nor proposition – that is, what is nonsense. Wittgenstein takes himself to have achieved this in the 'general propositional form': 'This is how things stand' (TLP 4.5). This is a variable whose values are *all* propositions: 'a description of the propositions of *any* sign language *whatsoever*' (TLP 4.5). If what *appears* to be a proposition does not, on analysis, conform to the form of all possible propositions, then it cannot really be a proposition and so does not really express a sense.[9]

Hacker on Nonsense as the Attempt to Say What May Only be Shown

Nonsense, then, is a failure to say how things stand in the world and thereby express a sense. Departing from what is said in the *Tractatus*, Hacker attempts to reconstruct Wittgenstein's understanding of nonsense in the light of TLP 6.54. This results in a more elaborate account of nonsense than is to be found in the text itself. Hacker distinguishes between 'overt' and 'covert' nonsense. Philosophy consists largely of the latter, which Hacker divides between a 'misleading' and an 'illuminating' variety.[10] The latter is Hacker's version of the notion of 'deep' nonsense that characterizes the orthodox account. Hacker maintains that, despite its nonsensicality, 'illuminating' nonsense is nonetheless able to convey insights into the relation between language and world that cannot be put into words. As such, it is able to transport its attentive recipient to what he calls a 'correct logical point of view'.[11] Once such a view is attained 'we shall apprehend what can and what cannot be said, and cease the futile attempt to say what can only be shown.'[12] This is Hacker's gloss on TLP 6.54. It is an attempt to explain how Wittgenstein could reject his text as nonsense, yet speak of understanding its point nonetheless and, as a result, of coming to 'see the world aright'.

Hacker maintains that philosophical nonsense arises when we 'try to say what can only be shown.'[13] What cannot be said is anything pertaining to the 'logical form' of the world: this shows itself in the logical form of legimately constructed propositions. Logical form is what language and world have in common: the fact that the logical form of language mirrors that of the world is what allows language to depict the world in the first place. But any attempt to frame propositions *about* logical form results in nonsense. As that by virtue of which propositions are able to picture reality, it cannot itself be pictured *in* propositions – or captured in thought: '[p]ropositions can represent the whole of reality, but they cannot represent what they must have in common with reality in order to be able to represent it – logical form' (TLP 4.12). This is impossible since it would entail that we 'station ourselves with propositions somewhere outside logic, that is to say outside the world' (TLP 4.12). Nonetheless, whilst it cannot be *represented by* language, Wittgenstein maintains that logical form *displays itself through* language:

> Propositions cannot represent logical form; it is mirrored in them.
> What finds its reflection in language, language cannot represent.
> What expresses *itself* in language, *we* cannot represent by means of language.
> Propositions *show* the logical form of reality.
> They display it. (TLP 4.121)

What is thereby shown '*cannot* be said' (TLP 4.1212). So what does Wittgenstein take himself to be doing when he *tells us* what cannot be said but only shows itself? Hacker maintains that he is trying to 'intimate', as it were, those 'ineffable metaphysical truths' about language and the world which, according to his own theory, cannot be directly communicated.[14] This leads to nonsense, but a peculiarly *philosophical* kind of nonsense. Hacker's distinction between two varieties of such nonsense is crucial at this point. 'Misleading' nonsense results from the failure to understand the principles of logical syntax, principles determining how the elements that make up propositions may be legitimately combined.[15] These rules for constructing propositions are governed by the combinatorial possibilities internal to the elements of which the world is composed, the form of which is mirrored by the logical form of the proposition. Hence the legitimacy of a proposition is ultimately determined by ontology. Any proposition put together in a manner that violates these principles will result in a nonsensical pseudo-proposition, the structure of which is not allowed given the way in which the elements in the world can be combined. 'Illuminating' nonsense also looks as if it is trying – but failing – to say something about the essence of the world. Yet it does so in a manner that guides the acute reader to 'apprehend' what is shown by other, *non-philosophical*, propositions: 'Illuminating nonsense will guide the attentive reader to apprehend what is shown by other propositions which do not purport to be philosophical; moreover it will intimate to those who grasp what is meant, it own illegitimacy.'[16]

Thus Hacker takes Wittgenstein's distinction between saying and showing to be a way of guarding against the counter-syntactic nonsense produced when we try to describe the logical form of language as revealed by philosophical analysis; and he takes Wittgenstein's self-destructing illuminating nonsense to be a means of guiding us see what such attempts *try* to say, and why they are bound to fail. This way of formulating the distinction between what can be shown but not said gives the impression that there is *something* we would like to say but cannot, because the rules of logical syntax prevent it. Wittgenstein's philosophy shows us both that we cannot say it, and why, and teaches us to abstain from our misguided attempts to do so. As Hacker puts it: 'The task of philosophy in this respect then is twofold, to bring one to see what shows itself, and to prevent one from the futile endeavour to say it.'[17] But to understand Hacker's point, it is necessary to consider what Wittgenstein says about 'formal' properties, concepts and relations immediately after formulating the distinction between saying and showing.

Hacker on 'Formal' Concepts, Properties and Relations

The formal properties and relations of propositions manifest the 'internal' properties and relations of the objects or states of affairs they depict, those that it is unthinkable the object or states of affairs should not possess (TLP 4.123). But that such properties obtain of a given object or state of affairs is not something that can be asserted in a well-formed proposition. Rather, this 'makes itself manifest' in propositions describing the object in question (TLP 4.122[iv]). Given that an object could not *not* possess any of its internal properties, it is as nonsensical to attribute a formal property to an object as it is to deny it. So, whilst it makes sense to say that 'the book is red', it is nonsensical to say that 'the book is an object', or that 'red is a colour'. Only the former is a genuine, bipolar, proposition saying something that might be otherwise; the other two try to say what could not possibly be otherwise – that is, what is shown by the former. Being an object and being coloured are internal properties of the concept 'book'. Hence, *that* the book 'is an object' or *that* red 'is a colour' would be *shown* through the signs by which 'red' and 'book' would be represented in an adequate notation: whereas a proposition attempting to *assert* that 'red is a colour' could not even be constructed. But it would also be unnecessary. For what it was trying to say would be apparent from the way in which a variable representing a colour-name could be combined with other signs to express a sense.

Hacker argues that such combinatorial possibilities are governed by the 'principles of logical syntax'.[18] In the last analysis, these principles are themselves dictated by the ontological type of simple objects picked out by logically proper names.[19] An object's 'ontological type' is determined, in turn, by its internal properties: 'for it is they that determine with what kinds of other objects it can combine to constitute a fact'.[20] Thus a spatial object, but not an auditory one, *could* combine with a colour. The basic idea, then, is that the logico-syntactical form of a name mirrors the ontological form of the object it picks out: 'just as the combinatorial possibilities of an object constitute its ontological type, so too the grammatical combinatorial possibilities of a name constitute its logico-syntactical category.'[21] Hence, the totality of ways in which it is possible to combine a sign, *x*, with other signs to form a legitimate proposition, where *x* represents a book, would show the formal properties of what it represented – that is, that a book *can* be so-coloured, so-extended, but not so-pitched. The combinatorial possibilities of the sign itself would *show* what the illegitimate proposition tried to *say* – namely, that a book 'is an object'.

One problem with this way of reading the *Tractatus* is that it gives rise to what appears to be a substantial realm of the ineffable; a realm in which there is something to be apprehended – that is, *that* the book is an object, *that* red is a colour – only it cannot be said, because the rules of logical syntax preclude its coherent formulation. Indeed, this much would seem to be built into the very idea that one *can* 'violate the logical syntax of language' by trying to say what may only be shown. Moreover, this would appear to rebound on the *Tractatus* itself: in so far as it attempts 'to say what may only be shown' it must, according to its own theory, be almost entirely nonsensical. Indeed, any attempt to explicate the saying/showing distinction puts the commentator in the similarly uncomfortable position of saying what can and what cannot be said, and why, thereby transgressing the fundamental requirement of meaningful discourse, the bipolarity of the proposition. Now, this might be taken to explain why Wittgenstein urges his reader to recognize his book as nonsense; but it begs a question as to what he could have hoped to achieve by writing such a book. Wittgenstein claims that climbing the ladder it provides will bring the reader to 'see the world aright'. But how could a book of nonsense – of all things – achieve such a feat?

'Illuminating' Nonsense – A Solution to TLP 6.54?

This is the question that leads Hacker to differentiate two varieties of philosophical nonsense – a distinction that he acknowledges is not only confusing but cannot be found in the text itself. The question is whether Hacker's distinction can do the work required of it. Hacker's claim, remember, is that illuminating nonsense 'guide[s] the attentive reader to apprehend what is shown by other propositions which do not purport to be philosophical' whilst 'intimat[ing], to those who grasp what is meant, its own illegitimacy'.[22] Hence, although the book is indeed nonsense, 'strictly speaking', once we have worked our way through it we may nonetheless emerge – if we 'grasp what is meant' – with an ability to discern what everyday propositions manifest about the essential nature of whatever they picture. Of course, we will not be able to say what this is, since 'what *can* be shown, *cannot* be said' (TLP 4.1212), but nor will we wish to, since we will have grasped the illegitimacy of trying to do so (this being what distinguishes illuminating from misleading nonsense). But how are we to 'grasp what is meant' when this is conveyed by nonsense and nonsense means nothing at all? However the problem is turned, Hacker's account of TLP 6.54 entails that *what nonsense 'means' be somehow understood.* But how can what is empty of content be understood? Nonsense makes no sense. It communicates nothing. If it communicates something it has a sense after all and therefore isn't nonsense.[23]

The problem with Hacker's idea that we may grasp what is meant by illuminating nonsense whilst being unable to say what is meant, is that the latter half of the formulation effectively denies what is asserted in the former. Why, if we can grasp what is meant, can we not say what we have grasped? Surely if we *have* grasped it, we must be cognisant of *what* we have grasped – and if we know what we have grasped, then why can't we *say* what we know?[24] Hacker's reason is that the syntactical rules for the construction of legitimate propositions preclude its coherent formulation. But this requires attributing to Wittgenstein the view that there is such a thing as having an insight which, *in principle*, cannot be expressed, when, according to the *Tractatus*, this would have to be wrong. The *Tractatus* holds that 'a thought is a proposition with a sense' (TLP 4). Only what would constitute a meaningful proposition could constitute a thought. But this does not mean that thought is *constrained* in some way; it means that a linguistic string without a sense does not present anything to think about. Should we find that whatever it is we want to say cannot be articulated in a meaningful proposition, then we cannot be trying to express a thought at all. Quite simply, whatever we *think* we have in mind here, it isn't a thought at all. Moreover, given that a thought is simply a 'propositional sign, applied and thought out' (TLP 3.5), it would be nonsensical to claim one had a thought in mind which couldn't be expressed. As far as the *Tractatus* is concerned, what can be thought is coextensive with what can be said. It is what is sayable. Everything that can be said can be thought, and nothing that can be thought is incapable of expression.

The question that this might be thought to leave open is whether there is anything out there, so to speak, which cannot be captured in either thought or language. Is there something – like getting hold of the ineffable metaphysical truths underlying our words – which we simply cannot do? Again, not according to the *Tractatus*. Moreover, *realizing this*, according to Conant and Diamond, is a precondition of understanding Wittgenstein's peculiar authorial strategy. One way of getting at what is at stake here is as follows. The *Tractatus* holds that the world may be completely described in meaningful propositions (TLP 4.26). Everything there is to say may be said. There is nothing our words are prevented from capturing in principle. It is not that doctrine of showing versus saying is false, but that it makes no sense. Or, rather, it is not a 'doctrine' at all. It cannot even be coherently stated. Every attempt to do so undoes itself. This is what Diamond means when she asserts that '[i]n so far as we grasp what Wittgenstein aims at, we see that the sentence-form[s] he uses come apart from his philosophical aim'.[25] The point is not to squint after what his sentences seem to be saying, but to realize that what looks like it is carrying meaning is actually vacuous. But the point is not solely deflationary. As a result of seeing this, we may also see that 'when we go in for philosophical thinking, the characteristic form of such thoughts is precisely that the sentence-forms we use come apart from what we have taken to be our aims'.[26] To respond by maintaining that, whilst the book's propositions may be 'strictly speaking' nonsense, they are nonetheless gesturing at truths that cannot be captured in words, is to block the very possibility of seeing what – according to Conant and Diamond – the *Tractatus* is designed to bring its readers to see. Moreover, it begs a question as to why the book takes the form that it does. From this perspective, if what is 'meant' by illuminating nonsense cannot be expressed in propositions, then it cannot be 'grasped' in thought. If we are unable to say what is meant – whether we realize it or not – we cannot have thought it either. Moreover, if it cannot be grasped in thought then, so far as the *Tractatus* is concerned, 'it' isn't out there to be grasped. Given the relation between language and world in the *Tractatus*, there is nothing for such nonsense to gesture towards.

This is the point at which we need to return to what Wittgenstein says in the preface. There, remarking on what he takes to be the point of book, he claims: 'Its whole meaning could be summed up somewhat as follows: What can be said at all can be said clearly; and whereof one cannot speak thereof one must be silent'.[27] Taken together with the identification of language and thought, and the assertion that what lies beyond the expression of thoughts will simply be nonsense (*wird einfach Unsinn sein*), this entails not only that Wittgenstein does not recognize, but that he precludes, distinctions of the kind Hacker's interpretation of the *Tractatus* requires. So, although illuminating nonsense is one obvious way of trying to make sense of the seemingly contradictory claims of TLP 6.54, not only is it internally inconsistent, but it cannot be rendered compatible with the text as a whole. This latter is important as it is would otherwise be credible to maintain that Wittgenstein had unwittingly burdened himself with a metaphysical picture of the precisely the kind his later work unceasingly sought to dissolve. The alternative, then, is to take Wittgenstein at his word when he specifies nonsense plain and simple. This is the crux of Conant and Diamond's response.

Conant and Diamond on the 'Frame' of Wittgenstein's *Tractatus*

According to Conant and Diamond, the reasons why Hacker's interpretation is incoherent can be found in the book's preface. But Hacker does not take Wittgenstein at his word. As a result, he attributes to Wittgenstein the very belief his preface diagnoses as philosophy's characteristic illusion. The belief that we can lever ourselves into a position – what Hacker calls the 'correct logical point of view' – from which we may spectate upon the world as if from outside it. Hacker maintains that this is the position from which the text claims to speak and to which it aims to bring its readers, despite noting that this 'could not possibly make sense'.[28] That Hacker reconstructs the text in a way that he concedes cannot make sense, and for reasons that the book itself adduces, suggests that he takes the *Tractatus* to be irredeemably confused. Yet it is Wittgenstein who provides Hacker with this argument: Wittgenstein who warns us against thinking we may peer beyond the limits of what can be said and thought; Wittgenstein who encourages us, in the preface and TLP 6.54, to understand him by recognizing his propositions as simple nonsense:

> [T]he aim of the book is to draw a limit to thought, or rather – not to thought but to the expression of thoughts; for in order to be able to draw a limit to thought, we should have to find both sides of the limit thinkable (i.e. we should have to be able to think what cannot be thought).
>
> It will therefore only be in language that the limit can be drawn, and what lies on the other side of the limit will simply be nonsense. (TLP preface, iii–iv)

To maintain that Wittgenstein wanted his readers to hold on to some ineffable metaphysical insights which his philosophical nonsense was trying to gesture towards therefore begs more questions than it answers. According to Conant and Diamond, it turns the *Tractatus* inside out by taking what the text *diagnoses* as an archetypical philosophical illusion as either a doctrine it espouses or a position to which it is implicitly committed. It is what Conant calls 'mistaking the target of the work for its doctrine'.[29] It attributes to the text what the text anticipates, and seeks to dispel, in its readers. If there is one conclusion that the Conant–Diamond reading suggests above all others, then, it is that we need a different perspective on the book as a whole, rather than a new interpretation of a particular stretch of remarks (although this is often true).[30] That is, a perspective through which it becomes possible to view the *Tractatus* as a 'work', in a sense normally reserved for literary works and works of art, conceived and written by a philosopher with a more acute sense of authorship than is generally found in the analytic canon – so raising a question as to how easily it can be accommodated within that canon, given its dominant self-understanding. Hence, although Conant and Diamond proceed by taking issue with text's orthodox commentators through a close reading of the book's 'frame' (essentially, though not exclusively, the preface and TLP 6.53–57), given that they see this frame as providing 'instructions for reading' the text as a whole, such a reorientation is what their account seeks to precipitate.[31] Hence, despite originating in little more than a subtle displacement of the orthodox interpretation of TLP 6.54, Conant and Diamond's reading ultimately stands the orthodox account of the text as a whole on its head.

'Throwing Away the Ladder': Conant and Diamond versus Hacker on TLP 6.54

Where Hacker's interpretation of TLP 6.54 requires that nonsense be understood, Conant and Diamond stress that Wittgenstein asks that *he*, and not his *propositions*, be understood. Both maintain that Hacker's determination to understand the book's propositions – despite their nonsensicality – rather than the author who tells his readers that they are nonsensical, is a refusal to 'throw away the ladder' and, so too, a refusal to 'see the world aright'.[32] Given that the debate pivots upon extrapolating the consequences of what Wittgenstein says at TLP 6.54, it will be helpful to remind ourselves at this point of what Wittgenstein himself says about 'throwing away the ladder':

> My propositions serve as elucidations in the following way: anyone who understands me eventually recognizes them as nonsensical, when he has used them – as steps – to climb up beyond them. (He must, so to speak, throw away the ladder after he has climbed up it.)
>
> He must transcend these propositions, and then he will see the world aright. (TLP 6.54)

Hacker's account of grasping what such nonsense, albeit obliquely, 'means' ends up treating nonsense as if it were propositional when, according to the *Tractatus*, nonsense is not a type of proposition – that is, a linguistic string with a sense – at all. Wittgenstein does not speak of 'viewing the world aright' by means of his propositions, but rather by 'transcending' them.[33] Of course, this is highly ambiguous: the more so since Wittgenstein likens his propositions to the steps of a ladder – suggesting that whatever we are to see has something to do with these propositions themselves. Here it is crucial to pay attention to the letter of Wittgenstein's text. Wittgenstein speaks of using his propositions to 'climb up beyond them'. And that, so far as Conant and Diamond are concerned, means getting beyond thinking one understands them, when, as nonsense, there is nothing to understand – except, that is, whatever can be made of the fact that they are nonsense.

Realizing this is fundamental. For only when Wittgenstein's propositions have been recognized as nonsense – that is, nonsense plain and simple – will they 'serve as elucidations'. This is why Conant and Diamond are so critical of interpretations that obscure this issue by invoking some version of what they believe to be deep nonsense. This move blocks the question that they believe the *Tractatus* is designed to make its readers pose from getting off the ground – that is, a question about the relation between a book that declares itself to be nonsense and what happens when we go in for philosophical thinking. From their perspective, Hacker's reading forecloses the possibility of grasping the point of Wittgenstein's book. This is what Diamond calls 'chickening out'.[34] Diamond defines 'chickening out' as 'pretend[ing] to throw away the ladder while standing firmly, or as firmly as one can, on it.'[35] Hacker leaves himself open to this rather unflattering description when he remarks: '[c]ertainly, Wittgenstein did not use the phrase 'illuminating nonsense'. What he said was that the propositions of the *Tractatus* elucidate by bringing whoever understands their author to recognize them as nonsensical'.[36] This is true. But because Hacker is more concerned to explain why Wittgenstein believes nonsense can convey ineffable truths than to reflect on what else might be at stake when a philosophical text flatly declares its own nonsensicality, he forecloses any possibility of understanding *Wittgenstein* which his own insight into TLP 6.54 might have provided.[37]

Conant and Diamond's response extrapolates the destructive consequences of this remark for Hacker's account as a whole. This is not to say that Conant and Diamond want to attribute a competing theory to Wittgenstein's text. Given that they take the point of Wittgenstein's text to devolve upon recognizing its nonsensicality, they want to demonstrate that it advances no theories at all.[38] To this end, they cite Wittgenstein's claim that philosophy is an activity of clarification, not a body of doctrine (TLP 4.112), and his prefatory warning that the *Tractatus* is not a book of doctrine (*Es ist…kein Lehrbuch*). From their point of view, all interpretations that construe the *Tractatus* as an attempt to communicate an ineffable theory refuse to throw away the final rung of the ladder. So, when Hacker claims that in the *Tractatus* 'the critique itself, the description of the limits of language, lies beyond the realm of what can be said. Language can no more describe its own essence than it can describe the limits of the world', he is merely reiterating the very 'doctrines' Wittgenstein claims are nonsensical, hence no doctrines at all, to support the conclusion that Wittgenstein was trying to gesture towards those 'ineffable metaphysical necessities' that elude direct communication.[39]

For Hacker, these are the ontological features of the world which *do* obtain, but cannot be *said* to obtain, because any attempt to do so 'violates the logical syntax of the language'.[40] This is what Diamond has in mind when she claims that Hacker fallaciously ascribes a 'realism of possibility' to the *Tractatus*.[41] Hacker believes that the 'rules of logical syntax', determining what can and what cannot be said, are governed by those possibilities that the ontological form of the world allows. Diamond claims that this amounts to burdening the text with the very conception of necessity that Wittgenstein's work consistently sought to contest – that is, 'necessity imagined as a fact,'[42] necessity conceived as a constraint that could have been otherwise had the world presented different ontological possibilities than those which, in fact, it does. This way of conceptualizing necessity and possibility pictures them as fixed one way rather than another, and hence as set within a wider realm of possible possibilities. What can and cannot be said depends on *which* necessities obtain, and the totality of what can be said is bordered by the space of what cannot – as things stand – be expressed (but which is there nonetheless).

This much is built into Hacker's notion of the 'correct logical point of view' – a point of view from which we are supposed to apprehend *both* what can and what cannot be said.[43] It also manifests itself in Hacker's distinction between illuminating and misleading nonsense. Given that both, on Hacker's account, 'violate the rules of logical syntax', the only real difference between them is the degree of cunning or self-consciousness with which they are yielded. The latter naively blunders into the confusion – that 'one can say things which can only be shown'[44] – which the former aspires to illuminate as confusion. But this entails not only that the purveyor of illuminating nonsense occupy a vantage point beyond the constraints that Hacker claims the structure of the world imposes upon what can be said – from where he 'apprehends' what cannot be said (or thought) – but also that, with consummate skill, he transgresses the rules governing what can be said in such a way as to bring them into view, and thereby raises his reader to a similar vantage point. As Diamond sees it, this is merely the semblance of a point of view that Wittgenstein imagines himself into in order to explode it from within: it is what the *Tractatus* is designed to show is only the *illusion* of a viewpoint.[45]

It is an illusion because it pictures the limits of what can be thought and said as if they were limitations, imagining that in plotting their outer edge we may somehow 'grasp' what lies beyond them. Yet it is to discourage the belief that we may 'think what cannot be thought' (TLP preface, iii) that the *Tractatus* tries to draw the limits of expression from within, by means of the general propositional form. Since this variable is the form of *all* propositions whatsoever, it encompasses all that can be said, excluding nothing. Hacker's account, by contrast, entails taking the *Tractatus* to harbour a notion of the totality of what can be said as limiting: it is therefore trying to bring what cannot be said into view by transgressing the boundaries of what can. But this gets the book back to front. It throws away the 'frame', which is designed to help us understand the book, rather than the book's contents, which we are asked to recognize as nonsense. Since it entails holding on to those contents, the orthodox account fails to completely throw away the ladder.

TLP 5.473–5.4733: Nonsense as the failure to Mean What We Say

By taking seriously what Wittgenstein says in his framing remarks, Conant and Diamond provide a very different account of what Wittgenstein was doing when he wrote a book which, they claim, is devoid of any theories whatsoever. Why anyone would want to write such a book and call it 'philosophy' would require explanation enough. In the case of the *Tractatus*, this peculiarity is compounded by a text that appears to be doing precisely what its preface claims is not there to be done, thereby encouraging its readers to do likewise, only to conclude by claiming that we will not have understood its author until we have realized his propositions are nonsensical. As Hacker's way of making sense of this appears flawed, we need to know more about how Wittgenstein construes nonsense in the *Tractatus*.

quote?

The relevant remarks run from TLP 5.473 to 5.4733: they are crucial to Conant and Diamond's attempt to oppose ascriptions of a deep conception of nonsense to the text. Hacker's version of this turns upon the idea that nonsense arises when we 'attempt to say what can only be shown'. This violates the logical syntax of language determining what combinations of signs can make sense. 'A is an object' is nonsense because it employs a formal concept ('object') as if it were a genuine concept in the attempt to say what is shown by the way the sign 'a' symbolizes in statements such as 'a is red'. The resulting pseudo-proposition is 'logically mal-formed'. It violates the 'rules of logical syntax' because only a material concept can be legitimately employed as a predicate. In no conceivable situation of use, and irrespective of any intention we might have to express a meaning, could these signs, combined in this way, make sense. Combinations of signs like this produce irredeemable nonsense. Now, whilst it is true that we can, and often do, combine words in ways that fail to express a thought, this is not, according to the *Tractatus*, because such would-be propositions *cannot* express a thought. It is because *we* have failed to mean anything by our use of these words in this way, because *we* have failed to determine what we mean by them. Here is what Wittgenstein says:

> Frege says every legitimately constructed proposition must have a sense. And I say that any possible proposition is legitimately constructed, and, if it has no sense, that can only be because we have failed to give a *meaning* to some of its constituents.
> (Even if we think that we have done so). (TLP 5.4733)

Leaving aside what Wittgenstein ascribes to Frege, he claims that every possible proposition, simply by virtue of being a proposition, is legitimately constructed.[46] A possible proposition is any proposition that conforms to the general propositional form, a variable taking all propositions whatsoever as its instances – the form of all that can be said. If it turned out that something we *believed* we were saying could not be translated into a logical notation, that would be because *we* had failed to express a sense by it, and hence had uttered nonsense. It would be a result of having not meant something determinate (or at least determinately indeterminate) by our words even though, to paraphrase Wittgenstein, we *thought* that we had. It would not be because what we took ourselves to be trying to say 'could not be said'. All that would be required to make this seemingly ineffable proposition meaningful – that is, to make it a genuine proposition – would be to determine what we wanted to say by it. Wittgenstein appends the following commentary to TLP 5.4733 making this clear:

> Thus the reason why 'Socrates is identical' says nothing is that we have not given *any adjectival* meaning to the word 'identical'. For when it appears as the sign of identity it symbolizes in an entirely different way – the signifying relation is a different one – therefore the symbols also are entirely different in the two cases: the two symbols have only the sign in common, and that is an accident. (TLP 5.4733)

The problem is that one sign symbolizes in two entirely different ways. Understanding what Wittgenstein is saying here presupposes an earlier set of remarks (TLP 3.3–3.33) in which he discusses the difference between a 'sign' and a 'symbol'. The latter refers to the sense of a proposition or, *given* that sense, the meaning of its constituent parts (TLP 3.3–3.31). The former refers to their perceptible qualities – that is, the words written down or spoken aloud (TLP 3.32). The problem with ordinary language is that the same signs are used with completely different meanings, or ways of 'symbolizing', depending up on the proposition in which they occur (TLP 3.323). The most obvious example of which is the little word 'is', which may function as the copula, the sign for identity, or of existence (TLP 3.323[ii]). But Wittgenstein makes it clear that the way to determine how a particular sign is symbolizing on a given occasion is to look at the context of use: 'In order to recognize the symbol by its sign we must observe how it is used with a sense' (TLP 3.326).[47] That is, we must ask what, in the circumstances, a speaker might mean by these words used in this way.

What is crucial when we do so is that only a proposition has a sense: individual words or phrases take their meaning from the contribution they make to the thought expressed by the whole: 'Only propositions have sense; only in the nexus of a proposition does a name have meaning' (TLP 3.3).[48] In order to determine how the word 'identical' is functioning in 'Socrates is identical' we only have to look at the proposition in which it occurs to realize that it is not functioning as a sign of identity since there is nothing specified for Socrates to be identical with.[49] The sentence has the same logical form as 'Socrates is mortal', and 'identical' is here predicated of Socrates in the same way as mortality might be. Our confusion stems from having no idea what 'identical' means when it is used like this – even if, as Wittgenstein remarks, we think that we do. The sentence fails to make sense not because it is 'illegitimately constructed', as Hacker would have it, for 'any possible proposition is legitimately constructed', but because we 'have failed to give a meaning to some of its constituents. (Even if we think that we have done so.)'. The remark in parentheses is crucial. We may *think* that we know what we mean by it, perhaps as a result of what 'identical' means elsewhere, or because of its similar construction to 'Socrates is mortal,' or because of some image of fitting Socrates into Socrates that occurs to the mind's eye, when in fact we don't. We are caught between all the meanings that 'identical', used like this, could be given, but without having determined which we want to give it. This would become apparent were we to try to translate it into a good notation.[50] As yet it means nothing. It is *plain* nonsense, nothing more. However, we only need to stipulate an adjectival meaning for 'identical' and we will have solved our lapse in meaning:

> Logic must look after itself.
> If a sign is possible, then it is also capable of signifying. Whatever is possible in logic is also permitted. (The reason why 'Socrates is identical' means nothing is that there is no property called 'identical'. The proposition is nonsensical because we have failed to make an arbitrary determination and not because the symbol, in itself, would be illegitimate). (TLP 5.473)

Pace Hacker, the only thing that differentiates a supposedly malformed proposition such as 'a is an object' from a meaningful proposition such as 'a is red' is that *we* have failed to assign a meaning to 'object' that fits its use in the former as a predicate-noun, and that is all that is wrong with it. Our confusion stems from what normally functions as a variable being used in a different way. Wittgenstein's point is not, as Hacker construes it, that 'a' really is an object only we cannot say so, but that 'object' is used here as a predicate-noun to which we have yet to assign a meaning. At best, we don't know what we want to say by it; at worst, no meaning we could assign it would satisfy us. The latter, according to Conant and Diamond, is endemic to philosophy.[51] It leads to the belief that the statement itself is gesturing towards something it never quite manages to say, because 'that' – whatever it is – would violate logical syntax. But statements don't express meanings – we do. Likewise it is us, not our statements, that fail to make sense. But there is nothing that prevents us from doing so in principle. Logical syntax does not prevent us from expressing what these words would say, could they be combined in this way. To maintain that it does is to contradict the *Tractatus* on two counts. First, it requires that a 'symbol, in itself would be illegitimate' (TLP 5.473); that these signs, combined in this way, *could not* express a sense. Whereas Wittgenstein maintains 'we cannot give a sign the wrong sense' (TLP 5.4732). Thinking that we can derives from the fallacy that the logical categories to which our signs belong do not fit together in this way. This contradicts the *Tractatus* on a second front. It is to imagine logic as a fact, a kind of cage, the bars of which we run up against, and sometimes even breach, in thought. But logic is not a cage; it makes thought possible. Given that the realms of what can be said and what can be thought are coextensive, it cannot retrospectively function as a limit on what we can and cannot say.[52] Wittgenstein even asserts that we cannot offend against it: 'logic looks after itself…in a certain sense we cannot even make mistakes in logic' (TLP 5.473).

Moreover, viewing syntax as a constraint obscures the distinction between symbol and sign. The connection between a sign and its meanings is arbitrary. What is important is how signs symbolize. This is what determines sense. But signs do not symbolize outside the context of a proposition with a sense (TLP 3.3), and how they then symbolize is determined by the contribution they make to the thought expressed by the whole. This has two consequences. First, given that signs do not symbolize *before* being combined to form a proposition with a sense, then, of themselves, they have no sense which could prevent their combination in any way at all. As Wittgenstein states: 'In logical syntax the meaning of a sign should never play a role. It must be possible to establish logical syntax without mentioning the *meaning* of a sign' (TLP 3.33). We cannot appeal to any meaning we believe the sign *already* has in order to say it cannot be used in a particular way. Thus, in itself, 'object' is just a sign waiting to be imbued with a meaning. Once it is, by being used in a meaningful proposition, that will determine the logical contribution it makes to the sense expressed by the whole. This is the second consequence. Words do not, of themselves – that is, outside the context of a meaningful proposition – have meanings. And any account, such as Hacker's, which claims that they do runs the risk of conflating the logical and the psychological.

Philosophical Elucidations: How to read the *Tractatus*

'Philosophy', properly understood, aims to clarify thoughts. It does not, the *Tractatus* informs us, advance theories. It is, echoing the preface, *keine Lehre* – not a body of doctrine. Rather, it is an *activity* that consists essentially of elucidation (*Erläuterungen*). Philosophy, and hence elucidation, attempts to make thoughts clear through the clarification of propositions:

> Philosophy aims at the logical clarification of thoughts.
> Philosophy is not a body of doctrine but an activity.
> A philosophical work consists essentially of elucidation.
> Philosophy does not result in 'philosophical propositions', but rather in the clarification of propositions.
> Without philosophy thoughts are, as it were, cloudy and indistinct: its task is to make them clear and give them sharp boundaries. (TLP 4.112)

The question is how we are to square this with the claim that the *Tractatus* consists almost entirely of nonsense. For how could a book of nonsense be intended to make thought clear by clarifying propositions? The *Tractatus* is, to say the least, a highly peculiar work. It opens with a preface in which its author anticipates not only that he won't be understood but that, if he is, only someone who has already thought along similar lines will understand what he is doing: and it concludes by saying we will not have understood its author until we realize that his propositions are nonsensical. Then they will 'serve as elucidations'. In between, we are given what appears to be an elaborate theory about what can and cannot be said and why. But the theory collapses. What's more, it is meant to collapse. Consider how Wittgenstein introduces the idea of logical form that grounds the distinction between saying and showing: 'In order to be able to represent logical form, we should have to be able to station ourselves with propositions somewhere outside logic, that is to say outside the world' (TLP 4.12). We should have to take up a position that is not there to be taken up. That, Wittgenstein suggests, is the reason for introducing the distinction between saying and showing in the first place. Of course, the irony is that in order to make anything of this distinction, this is the very position the book's orthodox commentators have to imagine they are able to adopt. Hence, the theory does not collapse because it is internally flawed: it collapses because it is nonsense. Moreover its author, as TLP 6.54 attests, is hardly unaware of the fact. What, then, are we to make of all this? Where – if anywhere – can such a book lead us?

We are now ready to throw away the ladder. Or rather, what once looked like a ladder leading to a correct view of the world *sub specie aternitatis* now looks like so much old wood; what once appeared to be an elaborate theory about what we can and cannot say has imploded under the weight of its own nonsensicality. Not only does this ladder not lead anywhere, it would not support our weight were we to stand on it. Yet, despite leaving us exactly where we were, without a grasp of a new theory to show for our exertions, we are – if the *Tractatus* has performed its function – where we were but in an *entirely different spirit*. Thus Wittgenstein's point is far from deflationary. For his book seeks nothing less than to transform the spirit in which we inhabit our lives, our world and our relations to others, and it seeks to do so by transforming our relation to our words. Like the *Philosophical Investigations*, the *Tractatus* is designed to show us how slovenly that relation can be, particularly when it comes to philosophy.[53] When Wittgenstein makes the same point in the *Investigations*, he expresses himself, as he does in the TLP preface and 6.54, in the first person, thereby addressing *us*, his readers, directly: 'My aim is: to teach you to pass from a piece of disguised nonsense to something that is patent nonsense'.[54] It is essential to what Wittgenstein is trying to achieve that his point be existentially appropriated by us, which is why this reading of the *Tractatus* may be called 'perfectionist'.[55] So this realization is not filed away as just one more philosophical datum amongst others – that is, that there are some people somewhere who sometimes mistake nonsense for sense – Wittgenstein tries to lure *us*, his readers, into mistaking nonsense for sense. In doing so, he is trying to get us to see for ourselves that whatever we may have *thought* we were thinking, we were not thinking anything.[56] This is why Wittgenstein's method can, despite appearances, be called clarificatory: it separates thought from the mere *appearance* of thought, sense from the mere *semblance* of sense, by focusing on the words by which we express ourselves (TLP 4.112).

Thus, to understand *Wittgenstein* (TLP 6.54) is in part to realize why his text takes the form that it does. Why, for example, the text does *not* adopt what it declares to be the 'only strictly correct method' in philosophy: that is, 'to say nothing except what can be said, i.e. the propositions of natural science – i.e. something that has nothing to do with philosophy – and then, whenever someone else wanted to say something metaphysical, to demonstrate to him that he had failed to give a meaning to certain signs in his propositions' (TLP 6.53). Generations of commentators have registered this fact, but failed to see that it is the book that puts this method in question, and not vice versa. For the book aims to bring *satisfaction*, which is what the method described in TLP 6.53 could never achieve.[57] To realize why the book takes the form that it does is thus to realize that its therapeutic ambitions are in keeping with the way it has been written and form it has been given. It is to realize that we, its readers, have been hoodwinked by nonsense masquerading as sense. And that is to grasp the authorial strategy which engenders a book that looks and reads like this.[58] Only then will Wittgenstein's propositions serve as 'elucidations' – by demonstrating that we are prone, when doing philosophy, to take nonsense for sense, to imagine we have made sense when we have failed to mean anything by our words.[59]

The insight to which the book aims to deliver us is thus an *insight into ourselves*. It is when we have come to this realization that we will have 'transcended' Wittgenstein's propositions. Then we will see the world 'aright' – that is, free from the illusion of a 'correct logical point of view'. Hence the silence to which the book's last proposition refers is not a pregnant silence guarding an ineffable truth. It is an injunction not to speak unless one means something by one's words, not because there is anything that we are debarred in principle from meaning, but because we are frequently given to meaning nothing at all. The book's final proposition – 'Whereof one cannot speak, thereof one must be silent'

– is thus little more than a tautology.[60] It reminds us that all there is to say can be said: if we have said, or tried to say, something we take to be 'unsayable', then we have actually said nothing at all. This is the realization that Wittgenstein wants to bring about in his readers when he claims that anyone who understands him will have recognized his propositions to be nonsense. And it is the form in which these propositions are delivered in the *Tractatus* that is designed to trigger this realization. Diamond describes this process as one in which Wittgenstein 'imagines himself' into the position from which the purveyor of philosophical nonsense thinks he speaks and, in so doing, leaves a trail that such a reader will think they can follow.[61] So, whilst all nonsense says nothing, not all nonsense *looks as if* it says nothing, and this is what Wittgenstein's strategy in the *Tractatus* depends on.[62] Given everything I have said about Hacker's explication of nonsense, it is important that this does not collapse back into a theory crediting nonsense with 'illuminating' powers – in the sense of harbouring a quasi-cognitive content. And indeed, contrary to Hacker's suggestion in his recent reply to Diamond that the idea of 'transitional' nonsense effectively reinstates the distinction she sets out to contest (between plain nonsense and 'important' nonsense), this idea does not require that *nonsense be understood*.[63] All it requires is that one bit of nonsense may bring its recipient to see that another, less self-evident, bit of nonsense *is* nonsense.[64] This is very different from claiming that nonsense may harbour an ineffable content. Hence elucidation, as Wittgenstein employs it in the *Tractatus*, entails fighting nonsense with nonsense in order to make nonsense apparent as such.[65] It is his early way of combating what he saw as the characteristic failure of philosophy, its tendency to take nonsense for sense. In the *Tractatus* this takes the form of increasing the pressure until the entire structure collapses under the weight of its own nonsensicality. In his later work it is the method employed, rather than the goal to which it aspires, that is fundamentally different.

Coda

Once the text of the *Tractatus* is seen from the perspective that Conant and Diamond's focus on its frame provides, much of what has bemused orthodox readers such as Hacker about a book which 'could not possibly make sense' falls away. For if Conant and Diamond are correct, the book is not even trying to make the kind of sense commentators like Hacker are looking for in its text. This is what is most impressive about the Conant–Diamond reading of the *Tractatus*: it grasps the *book*, rather than its contents, as a whole, without glossing over the difficulties that its apparent 'doctrines' have caused previous attempts to do so. In fact, it does so precisely by taking those difficulties seriously. As a result they succeed, rather remarkably, in showing this foundational work of analytic philosophy to be a literary achievement, leaving us in need of a different conception of tradition, and the affinities between texts, to which to return the book as it emerges. For this *Tractatus* is not just, nor even predominantly, a work of straightforward analytic argument. Indeed, whilst both have done much to show that it inherits a large part of its problematic from Frege, Conant in particular has tried to show that its way of dealing with that problematic – that is, *how* it inherits it, has affinities with Kierkegaard's pseudonymous project as well a certain aspects of Frege's work.[66] Given this, Conant's account of Wittgenstein's strategy of 'philosophical authorship' – that is, the way in which his text addresses its readers – calls for a conception of tradition that transcends the orthodoxies of the analytic–continental divide.

Hacker, not surprisingly, is unsympathetic to this endeavour, since not only does it run counter to his reading of the *Tractatus*, it takes issue with his view of the development of Wittgenstein's philosophy to which that reading gives rise and, so too, his view of the *Philosophical Investigations*. Indeed, since this paper was written Hacker has published a lengthy reply to Conant and Diamond. I cannot here do justice to the details of his response but I will give a brief overview.[67] Hacker's reply takes the form of what he calls a 'pincer movement,' drawing on internal evidence from the *Tractatus* itself and external evidence provided by Wittgenstein's remarks about it (dating from the time of its composition right through to the *Investigations*).[68] On the former front Hacker points out, not unreasonably, that the Conant–Diamond account relies on a highly selective reading of the book's remarks. More controversially, he argues that, in so far as their account relies on *any* remarks from the body of the text at all, it becomes 'methodologically inconsistent' since it must, by virtue of that fact, be predicated upon remarks they themselves claim to be nonsensical.[69] This point is not without bite: if nothing else it has forced Conant to finesse his interpretation.[70] Hence, until recently both Conant and Diamond were happy, despite leaning heavily on passages from the text (such as the remarks on nonsense (TLP 5.473–5.4733)), to make a hard and fast distinction between the book's frame and its contents – *all* of which they claimed were to be discarded. Hacker's reply has forced Conant to modify this: the frame is now said to consist of any remark – internal to the text or otherwise – that provides the framework necessary for Wittgenstein's elucidatory 'transitional' nonsense to perform its work.[71] This enlarged conception of the book's 'frame' is still said to provide meaningful instructions for reading without which the book could not hope to succeed; but the new stress on the framing remarks' *role*, at the expense of their *location*, represents a subtle shift from Conant's earlier formulations of his position.

That said, Hacker may be right in a more fundamental sense than that of the alleged 'methodological inconsistency.' This implies that Conant and Diamond are simply oblivious to the implications of their own procedure – that their interpretation, like the text itself on their account, ultimately saws off the branch on which it is sitting. But even if he is right, insofar as their account of why the text is nonsense is obliged to resort to nonsense in order to make its case and, hence, finally, to undo itself, his criticism would still fail to find its target. Or rather, it would fail to pursue the consequences of this fact to their conclusion: for the question which, if correct, his charge cannot but provoke – yet which Hacker does not ask – is what we, its readers, *are supposed make of this fact*.[72] What are we to make of a text that undermines itself and a commentary that explicates that fact, or so it seems, only at the cost of undermining itself? What are we to make of a text, and a commentary, that unravel in such an unseemly manner? We can either conclude that the Conant–Diamond reading is as internally confused as Hacker's account would render the *Tractatus*, or we may conclude that the content of their commentary is reflexively embodied in its form.[73] Hence, if their commentary *performatively undercuts itself*, this is because they understand themselves to be engaged in a similarly therapeutic activity to that which they attribute to Wittgenstein.[74] On such an interpretation, Hacker misses both.

On the second front, Hacker argues that Wittgenstein's work, immediately following his return to philosophy in 1929, is most plausibly understood as an attempt to patch up the crumbling theoretical edifice of the *Tractatus*.[75] He also draws attention to a variety of remarks Wittgenstein made between 1929 and the drafting of the *Philosophical Investigations* that appear to reflect his dissatisfaction with the claims advanced in the *Tractatus*. Hacker argues that if Conant and Diamond are correct that Wittgenstein never intended to provide a philosophical theory in the *Tractatus*, then it is impossible to make sense of Wittgenstein's remarks, or what he took himself to be doing, during this period. This external argument is one that Conant and Diamond have as yet done little to address. They might, of course, see these remarks – gleaned from lectures, letters and reported conversations, none of which were intended for publication by Wittgenstein – as largely circumstantial, and hence as inconclusive. Given their stress on strategies of 'philosophical authorship' in Wittgenstein's work – and hence with issues of formal *presentation* in that work's address to its anticipated readers – this would not be inconsistent. They might even seek to give a similar account of what Wittgenstein was doing in this period, given his sensitivity to the psychological difficulties of tackling illusions head-on, to their account of the *Tractatus*.[76] Alternatively, like Hacker, they might point to the implications of this being a period of transition in Wittgenstein's conception of philosophical method. Finally, they could draw on manuscript remarks more congenial to their own position, or give a contrasting interpretation of some of the remarks that Hacker himself relies on.[77] Be that as it may, the evidence Hacker marshals, circumstantial though it may be, has sufficient cumulative weight to require some response if it is not to cast a shadow over their reading of the *Tractatus* – which, internally at least, does a better job of grasping the *Tractatus* as a whole, and as a work, than Hacker's. This, then, is a point that Conant and Diamond still need to address.

But I will conclude with a few, very general, remarks about resources in the *Philosophical Investigations* that might be used support their interpretation of the *Tractatus*. For, although Conant and Diamond's account of the *Tractatus* has clear implications for how one views the relation between it and his later work, redescribing that relation in any detail would have taken me beyond the scope of this paper.[78] But one upshot of the Conant–Diamond reading discussed in this paper is that it makes the *Tractatus* an even greater failure than it is generally taken to be. This is ironic given how much they have done to demonstrate that it is not the *kind* of failure, to advance a metaphysical theory about the relation between language and the world, that it is widely regarded as. Yet, if their reading of the text is correct, the book fails to the degree that it succeeds. If the *Tractatus* is the elaborate ruse suggested by their account, then it is a highly successful one – so successful that it has failed to have the effect on its readers that their account suggests that Wittgenstein would have wanted. And that is why their reading provides an interesting slant on the later work. If one reads the *Tractatus* as they have suggested, it is a work of tremendous hubris. It presumes to know what its readers will believe and the fallacies to which they are prone: it presumes, in effect, to know its readers better than they know themselves. This might be because Wittgenstein views his readers largely in his own image, thereby refusing to acknowledge their 'separateness'[79] – suggesting that the temptation to philosophical theory-building which the Tractatus encourages is one that Wittgenstein himself felt – and encouraging in him the belief that he could, as it were, speak in advance, and without need of reply, and that, when he had spoken, the problems of philosophy would have been dissolved once and for all.[80] But not only is this a hugely overblown conception of what a single work of philosophy might achieve, it is flawed in two further respects. First, it is inadequate as a practice of 'therapy' – a term that remains remarkably ill-defined despite the frequency with which it is employed in debates about Wittgenstein – as anyone with any professional knowledge of how practices of therapy (of which there are many) actually work would attest. Second, and as a result, although the *Tractatus* clearly wants to bring about an ethical transformation in its readers with respect to their life with words, in respect of its *own* strategy of address, it is something of an ethical failure. In this regard the tireless responsiveness and moral fervour, to which Stanley Cavell has drawn attention in the writing of the *Investigations*, takes on new significance. From this perspective, it comes into focus as Wittgenstein's immanent critique – at the level of his later work's form and key methodological concepts – of the failure of the *method* of his early work and not, as is generally thought, its theories, to bring about a similarly therapeutic goal. To cite Hacker citing Wittgenstein:

> I might say: if the place I want to get to could only be reached by a ladder, I would give up trying to get there. For the place I really have to get to is a place I must already be at now. Anything that I might reach by climbing a ladder does not interest me.[81]

Notes

* I would like to thank to Stephen Mulhall and Rupert Read for their comments and correspondence on this paper in draft. I would also like to acknowledge Stephen Mulhall's supervision of my earlier work on Wittgenstein from which this paper derives.

1 Ludwig Wittgenstein, *Culture and Value*, ed. G.H. von Wright and H. Nyman, trans. P. Winch (Oxford: Blackwell, 1980), p. 18e.

2 I shall refer to the *Tractatus* throughout the text as 'TLP' followed by the relevant paragraph number.

3 P.M.S. Hacker, *Insight and Illusion: Themes in the Philosophy of Wittgenstein* (Oxford: Clarendon Press, 1986). The expressions in scare quotes are summaries of Hacker's position, as set out in this work (pp. 18–22 in particular), not direct quotations.

4 See Cora Diamond, *The Realistic Spirit: Wittgenstein, Philosophy and the Mind*, (Cambridge, MA: MIT Press, 1991), specifically, 'Throwing Away the Ladder: How to Read the *Tractatus*'. Although this paper, originally published in *Philosophy*, **63**, 1988, triggered the debate around TLP 6.54 with which I am concerned here, the Conant–Diamond approach to the *Tractatus* has precedents in the work of Peter Winch and Hide Ishiguro. See, for example, their respective papers in Peter Winch (ed.), *Studies in the Philosophy of Wittgenstein* (London: Routledge, 1969).

5 For an interpretation of this kind see Norman Malcolm's *Nothing is Hidden: Wittgenstein's Criticism of his Early Thought* (Oxford: Blackwell, 1986).

6 It is important to note that, according to the *Tractatus*, only a linguistic string with a sense (including one structured so as to cancel itself out) is a proposition. Thus, unlike senseless propositions that say nothing but nonetheless show something (that is, *that* they say nothing) nonsense *neither says nor shows anything*. Nonsense, in other words, is *not propositional.*

7 There is no such thing as 'thinking what cannot be said', for 'a thought is [simply] a proposition with a sense' (TLP 4).

8 Hence, any thought that appears to fall outside it can only be a *semblance* of thought, for there is no thought that cannot be thought.

9 On the centrality of the general propositional form to Wittgenstein's aims in the *Tractatus* see Cora Diamond's 'Ethics, Imagination and the Method of Wittgenstein's *Tractatus*', in R. Heinrich and H. Vetter (eds), *Bilder der Philosophie, Wiener Reihe*, **5** (Vienna: Oldenbourg, 1991), section 1. This is reprinted in A. Crary and R. Read (eds), *The New Wittgenstein*, (London: Routledge, 2000).

10 Hacker, *Insight and Illusion*, *op. cit.*, n. 3, pp. 18–19. See also pp.18–26.

11 Ibid., pp. 22 and 26.

12 Ibid., p. 22.

13 Ibid., p. 18ff.

14 Defending the *Tractatus* from its appropriation by the Vienna Circle, Hacker claims 'Wittgenstein did think, when he wrote the Tractatus, that there were ineffable metaphysical necessities', (ibid., p. 54). Peter Geach also maintains that language manifests 'necessities founded deep in the nature of things' in 'Saying and Showing in Frege and Wittgenstein', in J. Hintikka (ed.), *Essays on Wittgenstein in Honour of G.H. von Wright*, *Acta Philosophica Fennica*, **28** (Amsterdam: North Holland, 1976), pp. 54–70.

15 'Philosophers try to say what may only be shown, and what they say, being nonsense, does not even show what they try to say' (Hacker, *Insight and Illusion*, *op. cit.*, n. 3, p. 18).

16 Ibid., pp. 18–19.

17 Ibid., p. 19.

18 Ibid., pp. 19–22.

19 Ibid., p. 20. The relevant background on the picture theory and the apparent 'logical atomism' of the *Tractatus* may be found in Hacker's *Insight and Illusion*, *op. cit.*, n. 3, or Malcolm's *Nothing is Hidden*, op. cit., n. 5, or Anthony Kenny's Wittgenstein (Harmondsworth: Penguin, 1973). For a contrasting view, see Hide Ishiguro's 'Use and Reference of Names' in Winch, *Studies in the Philosophy of Wittgenstein* (*op. cit.*, n. 4), p. 35.

20 Hacker, *Insight and Illusion*, op. cit., n. 3, pp. 19–20.

21 Ibid., p. 20.

22 Ibid., p. 18.

23 One of the ironies of the debate between Hacker and Diamond is that Diamond has done far more than Hacker to show how nonsense can be 'illuminating' *despite being devoid of cognitive content*. Her claim is not that one may *understand* nonsense, but rather that nonsense can carry the *semblance* of an argument that allows one to *think* one is following it and so can deliver a reader from the beginning to the end of a text. See Diamond, *The Realistic Spirit*, *op. cit.*, n. 4, Introduction II, p. 34; the account of 'Riddle-Reasoning' in ch. 10; and the discussion of imaginatively entering into the nonsense another takes for sense in 'Ethics, Imagination and the Method', *op. cit.*, n. 9.

24 Hacker claims that we may 'apprehend' that there is something which such nonsense is trying, but failing, to say. On this point see James Conant's 'The Search for Logically Alien Thought: Descartes, Kant, Frege and the Tractatus', *Philosophical Topics*, **20** (1), 1991, pp. 101–80, especially pp. 136–46. This is an analysis of what is wrong with the idea that we may somehow 'apprehend' what we cannot 'comprehend'. Conant argues that this ultimately relies on the incoherent and psychologistic notion that we can *think that* there is something we cannot think – that is, that there is something which *we* cannot think because it eludes *our* thought (and not because it simply isn't a thought at all).

25 Diamond, 'Ethics, Imagination and the Method', *op. cit.*, n. 9, p. 198.

26 Ibid., 198–99.

27 This is Ogden's translation of the German '*Was sich überhaupt sagen lässt, lässt sich klar sagen; und wovon man nicht reden kann, darüber muss man schweigen*' (TLP, preface, ii). Pears and McGuinness render this as: 'what can be said at all can be said clearly, and what we cannot talk about we must pass over in silence'. I have used Ogden since Pears and McGuinness tend to load the translation with a greater sense of there being *something* substantive which is 'passed over' than is present in the German.

28 Hacker, *Insight and Illusion*, *op. cit.*, n. 3, p. 25.

29 See James Conant's 'Elucidation and Nonsense in Frege and Early Wittgenstein' in Crary and Read, *The New Wittgenstein*, *op. cit.*, n. 9, p. 177.

30 One danger of speaking of *the* 'Conant–Diamond reading' is that it risks collapsing the differences between them. Whilst they are substantially in agreement, Diamond's interests in the *Tractatus* might be characterized as more ethical than Conant's, and Conant more focused on philosophical authorship than Diamond. Nonetheless, I shall employ this shorthand unless a clear difference would be obscured as a result.

31 On this 'frame' see Diamond's 'Ethics, Imagination and the Method', *op. cit.*, n. 9, pp. 55–59; and James Conant's 'Throwing Away the Top of the Ladder', *Yale Review*, **79**, 1991, especially pp. 344–47.

32 See Diamond, 'Ethics, Imagination and the Method', *op. cit.*, n. 9, pp. 194–98.

33 Cora Diamond has pointed out to me, in correspondence, that this now standard translation of '*überwinden*' in TLP 6.54 (*Er muss diese Sätze überwinden*) as 'He must 'transcend' these propositions' is inadequate in so far as it robs the word of any hint of struggle (and hence of any suggestion of a need to struggle with Wittgenstein's propositions). This would be apparent were a more literal, everyday translation of '*überwinden*', such as 'overcome', employed.

34 'Throwing Away the Ladder: How to Read the *Tractatus*', originally appeared in *Philosophy*, **63**, 1988. This paper now forms Chapter 6 of Diamond, *The Realistic Spirit*, *op. cit.*, n. 4. All references will be to the latter.

35 Diamond, 'Ethics, Imagination and the Method', *op. cit.*, n. 9, p. 194; see pp. 194–201 in particular.

36 Hacker, *Insight and Illusion*, *op. cit.*, n. 3, p. 26.

37 After this remark Hacker immediately asserts – presumably expressing what he takes Wittgenstein to have believed: 'the pseudo-propositions of the *Tractatus* lead one to see the world aright, from a correct logical point of view' (ibid., p. 26).
38 This already suggests the kind of consequences that Conant and Diamond's reading of the *Tractatus* will have for reconsidering its relation to the *Investigations*. See, for example, Ludwig Wittgenstein, *Philosophical Investigations*, ed. R. Rhees and G.E.M Anscombe, trans. G.E.M Anscombe, 2nd edn (Oxford: Blackwell, 1958), § 109: 'We may not advance any kind of theory … We must do away with all explanation and description alone must take its place …. The problems are [dis]solved, not by giving new information, but by arranging what we have always known.' Whilst the method described is clearly very different, the aspiration to *dissolve* philosophical problems by making our relation to our words perspicuous is substantially in accord with the *Tractatus*.
39 Hacker, *Insight and Illusion*, *op. cit.*, n. 3, pp. 23 and 54.
40 Thus Hacker writes: 'Metaphysical necessity is *ineffable*, but is *shown* by empirical propositions, namely by *features* (logico-syntactical forms) of their constituent symbols' (ibid., p. 51).
41 Diamond, *The Realistic Spirit*, *op. cit.*, n. 4, p. 194.
42 Ibid., p. 195.
43 Ibid., pp. 194–96.
44 Hacker, *Insight and Illusion*, *op. cit.*, n. 3, p. 19.
45 Diamond, *The Realistic Spirit*, *op. cit.*, n. 4, p. 195. See also 'Ethics, Imagination and the Method', *op. cit.*, n. 9, in which Diamond discusses this notion of imagining oneself into the position which the utterer of philosophical nonsense believes they occupy and from which, as it were, they try to peer back at the world as if from outside it.
46 Charting Wittgenstein's relation to Frege is beyond the scope of this paper. The most relevant orthodox account of this is Geach's 'Saying and Showing', *op. cit.*, n. 14. For the Conant and Diamond reply see Diamond, *The Realistic Spirit*, *op. cit.*, n. 4, chs 2–4 and Conant's 'Elucidation and Nonsense', *op. cit.*, n. 29.
47 Of course, this is one slogan traditionally employed to express Wittgenstein's views on meaning in the *Investigations*, further indicating a generally underplayed concordance between the two works.
48 This is a reworking of Frege's 'context principle' from the *Grundlagen*: 'never […] ask for the meaning of a word in isolation, but only in the context of a proposition.' See Gottlob Frege, *The Foundations of Arithmetic*, trans. J.L. Austin (Evanston, Ill.: Northwestern University Press, 1980), p. x. In the *Tractatus* this entails that the meaning of words cannot be determined in advance of the sense of the proposition as a whole in which they occur. For an early, 'anti-orthodox', discussion of the relation of Wittgenstein's conception of propositional sense and the reference of names in the Tractatus to Frege's context principle see Ishiguro's 'Use and Reference of Names', *op. cit.*, n. 19, pp. 20–25.
49 If we wanted to know what, if anything, it does mean we would have to begin by considering the context that prompted a speaker to give voice to it. That is, what might someone uttering these words, combined in this way, mean in this situation?
50 For an account of how logical notation becomes, in Wittgenstein, a tool of philosophical therapy which may be used to free ourselves from the hold that sentences, by which we have failed to mean anything, have upon us, see James Conant's 'Must We Show What We Cannot Say?', in R. Fleming and M. Payne (eds), *The Senses of Stanley Cavell*, Bucknell Review (Lewisburg, PA: Bucknell University Press, 1989), pp. 258–60.
51 See, for example, Diamond, *The Realistic Spirit*, *op. cit.*, n. 4, pp. 195–98.
52 See James Conant's genealogy of this conception of logic in 'The Search for Logically Alien Thought', *op. cit.*, n. 24.

53 The realm in which, according to Wittgenstein's *Philosophical Investigations, op. cit.*, n. 38, problems arise when our language 'idles' (§ 132) or our words have been divorced from any context of *significant* use (§ 116). Or, as § 38 puts it: 'philosophical problems arise when language goes *on holiday*'.
54 Ibid., § 464.
55 Thus one can see Conant and Diamond's emphasis on *how to read* the *Tractatus* in the light of Cavell's writings on perfectionism and the kind of demand for transformation which perfectionist texts make upon their readers. This theme pervades Cavell's work, but see, for example, his 'Existentialism and Analytic Philosophy', in *Themes Out of School* (San Francisco: North Point Press, 1984), especially § 3.
56 On this point see Conant's 'The Search for Logically Alien Thought', *op. cit.*, n. 24, pp. 136–47.
57 Hence the aim of the *Tractatus* – to bring one's philosophical interlocutors, and the philosopher in oneself, *satisfaction* – is consistent with that of the Investigations. See, for example, Wittgenstein, *Philosophical Investigations, op. cit.*, n. 38, § 133: 'The real discovery is the one that makes me capable of stopping doing philosophy when I want to. – The one that gives philosophy peace, so that it is no longer tormented by questions which bring *itself* in question'.
58 See Conant's 'Putting Two and Two Together: Kierkegaard, Wittgenstein and the Point of View for Their Work as Authors', in T. Tessin and M. von der Ruhr (eds), *Philosophy and the Grammar of Religious Belief* (London: St Martin's Press, 1995), for a comparison of the authorial strategy evinced by Kierkegaard's pseudonymous works and Wittgenstein's *Tractatus*. In particular Conant shows how what Kierkegaard has to say on the delicate handling required to disabuse a reader of an illusory point of view may be fruitfully applied to the strategy of the *Tractatus*.
59 Given that, for Wittgenstein, this means working in the tradition inaugurated by Frege and Russell's attempt to construct an ideal philosophical notation, his target is the illusory self-understanding to which we are always at risk of succumbing in such an undertaking. That not only is everyday language somehow defective but that were we able to achieve an ideal notation then the true logical form of the world would finally be looking back at us.
60 Whether or not we take it to be a tautology, according to Conant, will depend largely on *us* in the final analysis: that is, on the kind of sense we are likely to think we can perceive in it when trying to make sense of Wittgenstein's book. See Conant's 'Elucidation and Nonsense', *op. cit.*, n. 29, fn. 102, pp. 216–17.
61 See n. 23 above.
62 For a similar point see fn. 78 of Conant's 'Elucidation and Nonsense', *op. cit.*, n. 29, p. 209.
63 See Hacker's 'Was He Trying to Whistle It?', in Crary and Read, *The New Wittgenstein, op. cit.*, n. 9; and Conant's 'Elucidation and Nonsense', *op. cit.*, n. 29, p. 196.
64 Again, compare this to Wittgenstein's declaration at *Philosophical Investigations, op. cit.*, n. 38, § 464: 'My aim is: to teach you to pass from a piece of disguised nonsense to something that is patent nonsense'.
65 Conant brings this out in 'The Search for Logically Alien Thought', *op. cit.*, n. 24, and 'Elucidation and Nonsense', *op. cit.*, n. 29. In both, Conant traces the minutae of Wittgenstein's response to the tensions he found in Frege's thinking. The latter article in particular suggests that Wittgenstein resolves a tension between a 'deep' and an 'austere' conception of both nonsense and elucidation in favour of an austere conception in his own work. Following Diamond, Conant singles out Frege's debate with Kerry in 'On Concept and Object' (in P.T. Geach and M. Black (eds), *Translations from the Philosophical Writings of Gottlob Frege* (Oxford: Blackwell, 1984)) as a paper in which Frege's

conflicting uses of these terms can be clearly seen. On this point see also the early chapters in Diamond's *The Realistic Spirit*, *op. cit.*, n. 4, and Joan Weiner, *Frege in Perspective* (Ithaca, NY: Cornell University Press, 1991), especially ch. 6.

66 See Conant, 'Must We Show What We Cannot Say?', *op. cit.*, n. 50, and 'Putting Two and Two Together', *op. cit.*, n. 58, and 'Kierkegaard, Wittgenstein and Nonsense', in T. Cohen, P. Guyer, and H. Putman (eds), *Pursuits of Reason* (Lubbock: Texas Technical University Press, 1992). These parallels between the strategies of philosophical authorship evinced by Wittgenstein and Kierkegaard – specifically, whether or not we may take their statements at face value – open on to questions about the extent to which such concerns about philosophical method and authorship also figure prominently in the writing of Frege. See n. 65 above and n. 72 below.

67 See Hacker, 'Was He Trying to Whistle It?' *op. cit.*, n. 63. Hacker's response is collected, along with essays by Conant and Diamond and related essays contesting the standard portrayal of Wittgenstein's philosophy in Crary and Read, *The New Wittgenstein*, *op. cit.*, n. 9. This collection is the best place to turn for an overview of this debate. Unfortunately it appeared after this paper was written: all references to Conant's 'Elucidation and Nonsense in Early Frege and Wittgenstein' above were originally taken from an unpublished draft of that paper entitled 'Elucidation in Early Frege and Wittgenstein'. There is a second instalment of Hacker's response entitled 'When the Whistling Had to Stop' forthcoming.

68 Crary and Read, *The New Wittgenstein*, *op. cit.*, n. 9, p. 360.

69 Ibid., pp. 360–62.

70 Evidence of this process emerges in the footnotes to Conant's 'Elucidation and Nonsense', *op. cit.*, n. 29. See, for example, fn. 68, 96 and especially 102. Lynette Reid raises similar worries to Hacker's about 'methodological inconsistency' in 'Wittgenstein's Ladder: The Tractatus and Nonsense', *Philosophical Investigations*, **21** (2), pp. 97–151, to which Conant may also be responding.

71 See Conant's 'Elucidation and Nonsense', *op. cit.*, n. 29, fn. 102, pp. 216–17. This adjustment has one further, unanticipated, benefit for the Conant–Diamond reading – namely, that if the *entire* body of the text is no longer construed as ironic 'transitional' nonsense then, presumably, the entire frame need not, *mutandis mutandi*, be construed as straightforwardly sensical. This creates greater scope to deal with remarks in the prefatory frame that might otherwise present their account with obvious difficulties – such as Wittgenstein's claim that 'the *truth* of the thoughts communicated here seems to me unassailable and definitive' (TLP, preface, viii). For an account that already departs from a hard and fast demarcation between a sensical frame and nonsensical contents see J. Floyd's 'The Uncaptive Eye: Solipsism in Wittgenstein's *Tractaus*', in L.S. Rouner (ed.), *Loneliness*, Boston University Studies in Philosophy and Religion (Notre Dame: University of Notre Dame Press, 1998), p. 87.

72 That Hacker does not ask this question is all the more surprising given the weight that Conant and Diamond put on Frege's paradoxical nonsensical response to Kerry: 'The concept horse is not a concept.' See, for example, Conant's 'Elucidation and Nonsense', *op. cit.*, n. 29, especially pp. 185–89.

73 Hacker is ready to accept the consequences of his interpretation for Wittgenstein's text, claiming that there is no reason to assume that the *Tractatus* is an internally consistent work. See his 'Was he Trying to Whistle it?', *op. cit.*, n. 63, p. 370. One response to this is to see the *Tractatus* as Wittgenstein's response to his *own* temptations to theory-building (of the kind evinced in Ludwig Wittgenstein, *The Notebooks, 1914–1916*, ed. G.H. von Wright and G.E.M. Anscombe, trans. G.E.M. Anscombe (Oxford: Blackwell, 1969) and the more

Schopenhauerian aspects of the *Tractatus*), requiring that the text contain both what appears to be an elaborate theory and the seeds of that theory's destruction.

74 On this point see p. 191 and fn. 77 to Conant's 'Elucidation and Nonsense', *op. cit.*, n. 29 and the references to Cora Diamond's idea of imaginatively entering into another's nonsense and her conception of 'riddle reasoning' in n. 23 above.

75 To support this claim, Hacker cites 'Some Remarks on Logical Form' in Ludwig Wittgenstein, *Philosophical Occasions 1912–1951*, ed. J. Klagge and A. Nordmann (Indianapolis: Hackett, 1993) and Desmond Lee's notes of Wittgenstein's Cambridge Lectures from c.1930–31 in Wittgenstein, *Wittgenstein's Lectures, Cambridge 1930–32, From the Notes of John King and Desmond Lee*, ed. D. Lee (Oxford: Blackwell, 1980). See also 'The Colour Exclusion Problem' and 'Dismantling the *Tractatus*' in Hacker, *Insight and Illusion*, *op. cit.*, n. 3, pp. 108–20.

76 This is the point at which Conant, in particular, has had recourse to Kierkegaard's pseudonymous project, particularly as set out in 'Putting Two and Two Together', *op. cit.*, n. 58. See Søren A. Kierkegaard, *The Point of View for My Work as an Author*, trans. W. Lowrie (New York: Harper & Row, 1962). But one hermeneutic question this appeal raises, is why we should accept *this* statement as either sincere or straightforward when such hermeneutic scepticism is levelled at the claims advanced elsewhere in his work. That is, whether, for all Conant's intense formal scrutiny of philosophical works, this appeal to the author's real – that is, truly intended – meaning falls foul of a kind of 'intentional fallacy'.

77 For example, Hacker concludes his account with the following citation from *Culture and Value*: 'I might say: if the place I want to get to could only be reached by a ladder, I would give up trying to get there. For the place I really have to reach is a place I must already be at now. Anything that I might reach by climbing a ladder does not interest me' (in *The New Wittgenstein*, *op. cit.*, n. 9, p. 382). Hacker takes this rejection of the ladder metaphor to support his understanding of Wittgenstein's later view of the *Tractatus*. But it is equally, if not more, compatible with Conant and Diamond's reading. For they could respond that there is, in the last analysis, no ladder in the *Tractatus* – just the *appearance* of one. Hence, when the reader finally throws the ladder away they are left where they were before but in a different spirit – that is, without a certain kind of hankering. This thought complements many of the remarks in the *Investigations* about bringing philosophy peace. Examples such as this, which can be read in two entirely different ways depending on the interpretation guiding the reading, suggest that what Conant and Diamond are after, above all, is a kind of aspect-shift with respect to Wittgenstein's philosophy as a whole.

78 But see, on this point, the essays collected in Part I, Crary and Read, *The New Wittgenstein*, *op. cit.*, n. 9.

79 This is to adapt a term of Cavell's. See, for example, part IV of Stanley Cavell, *The Claim of Reason: Wittgenstein, Skepticism, Morality and Tragedy* (Oxford: Oxford University Press, 1979).

80 This may shed further light on a remark from the book's preface which ought to trouble supporters of the Conant–Diamond reading: 'the *truth* of the thoughts communicated here seems to me unassailable and definitive. I am, therefore, of the opinion that the problems have in essentials been finally resolved.' See n. 71 above.

81 Wittgenstein, *Culture and Value*, *op. cit.*, n. 1, p. 7e.

- Addresses the dispute btw. H & C+D.
- A good overview of H's position & how its diff. parts can be seen as hanging together.
  - Explains how H is in the grip of the illusion of seeing both sides of the limit.
- Perfectionism + Cavell (115)
- Probl: 116. TLP as a literary achievement.
- 117: calls for transcending anal.- cont. divide.
- The disc. of H's reply is not very successful.
- Does not recogn. the role of persp. pres. in C&D.
- Doubtful acc. of TLP's failure. (119)

Chapter 6

More Making Sense of Nonsense: From Logical Form to Forms of Life*

Daniel D. Hutto

> When I was in Norway during the year 1913–1914 I had some thought of my own, or so at least it seems to me now. I mean I have the impression that at the time I brought to life new movements in thinking (but perhaps I am mistaken). Whereas now I seem just to apply old ones.[1]

Introduction

A familiar way to read the Wittgenstein corpus is to see it as split into two periods during which two radically different accounts of the nature of language are advanced. Such great emphasis is often placed on this shift that it is common to speak of 'two Wittgensteins', the early and the late. On this reading, Wittgenstein's later writings are best understood as a reaction to, even a straightforward rejection of, his early work, which culminated in the *Tractatus Logico-Philosophicus*.[2] Crudely, he is alleged to have radically revised his understanding of how language operates by abandoning the 'picture theory of meaning' and its attendant metaphysics and semantics and advancing, in its place, the idea that 'meaning is use'. Amongst other things, this reading often inspires the thought that his philosophy underwent an important shift from realism to anti-realism, when his views changed. For convenience, I will class such readings of this sort as 'doctrinal' interpretations, since they promote the idea that in order to understand Wittgenstein's philosophy one must primarily focus on his theories about language.

This standard form of interpretation has recently been challenged by those who seek to demonstrate that it cannot be easily made to fit with Wittgenstein's claim that he was not engaged in any form of philosophical theorizing at all. For example, in a series of important papers, James Conant objects that the doctrinal interpretation cannot accommodate the remark that philosophy, '...is not a body of doctrine'.[3] On such grounds, both he and Cora Diamond argue convincingly that any simple doctrinal interpretation fails, '...to take seriously what Wittgenstein says about philosophy itself'.[4] Furthermore, these detractors emphasize that such claims were made during both periods. But if there are no theories or doctrines advanced, even in the *Tractatus,* then familiar talk of the 'picture theory of meaning' and the 'doctrine of showing' is wrongheaded. For the fact is that it is not possible to make sense of such labels unless it is also accepted that the book contains, '...numerous doctrines which Wittgenstein holds cannot be put into words'.[5] Diamond regards such acceptance as a 'chickening out' response brought on by a failure of nerve, poor interpretation or both.

Instead, they bid us to focus on Wittgenstein's self-avowed method of clarification by which sense and nonsense are to be distinguished. They ask us to take heed of his warning against what would otherwise be a natural misreading. He explicitly says:

> Philosophy aims at the logical clarification of thought.
> Philosophy is not a body of doctrine but an activity.
> A philosophical work consists essentially of elucidations. (TLP 4.112)[6]

In line with this, we are urged to adopt a very austere reading of the *Tractatus* that sees it in terms of a therapeutic project, with the purpose of 'working on the reader'. Wittgenstein's aim remained constant during his early and late periods. It was always, '...to take the reader from a piece of disguised nonsense to a piece of undisguised nonsense'.[7] Read in this light one can, '...insist upon a greater continuity in his work than most previous commentators have allowed'.[8] Yet, neither Conant nor Diamond deny that there is '...an equally significant discontinuity in the form of the investigation through which this aim is prosecuted'.[9]

There is much to recommend this reading – both as a corrective and especially as a means of getting us to focus on otherwise neglected issues in Wittgenstein studies such as the style of his authorship.[10] Moreover, it makes the study of his early writings as important as his later ones if we seek to understand properly his views on the nature and end of philosophy. However, in its extreme form, the therapeutic interpretation is implausible and threatens to obscure some important aspects of the development of Wittgenstein's thought. In this respect, used as the sole means of reading Wittgenstein, it can be as distorting as the more popular doctrinal interpretation.

As I have said, Diamond holds that to adopt this reading properly we must 'resolutely' give up the idea that Wittgenstein ever attempted to advocate any doctrines (cf. TLP 4.112). She recommends that to the extent that any of the remarks in the *Tractatus* appear to gesture towards something metaphysically external to language they should be read in an ironic sense.[11] In her view, '...the notion of something true to reality but not sayably true is to be used only with the awareness that it itself belongs to what has to be thrown away'.[12] However, coming to terms with the nature of Wittgenstein's message in the *Tractatus* is tricky. In my view we must distinguish the claim that the *Tractatus* is advancing some kind of 'theory' from the claim that it makes genuine assumptions about the nature of language, in a way that Diamond's reading hold it does not. Put simply, I shall argue that we ought to adopt an anti-metaphysical reading of the work, while firmly retaining the view that it does make assumptions about the nature of language that it ought not. This is necessary if we want to accept that the book has an internal tension and I will be arguing that it is only by abandoning the ironic reading and doing so that it is possible to make best interpretative sense of it.

Let us begin by focusing on what is right about the therapeutic reading, by considering the nature of Wittgenstein's clarificatory project and the extent to which he was successful in prosecuting it.

Philosophical Nonsense

To many, the *Tractatus* appears to be a strange, escalating series of philosophical pronouncements. As Brockhaus suggests, it is usual for it to be viewed as an undefended presentation of semi-independent metaphysical theses. He writes:

> Compressed, presented *ex cathedra* without argument and in a curious vatic voice, it offers an accelerating series of remarks on the world, logic, and the essence of language, suddenly and quite mysteriously blossoming into cryptic claims about the will, ethics, 'God' and 'The Mystical'.[13]

Yet I claim that rather than embodying a set of dogmatically advanced statements, its remarks are meant to provide a kind of philosophical therapy. To this extent I agree with Conant and Diamond, but I do not accept that these remarks are advanced in an ironic spirit or with the intent that they should be 'thrown away'. They are supposed to get us to see what ought to be obvious to us all. To use the language of the later writings, they are supposed to be mere reminders of what we ought already to know.[14] They are supposed to be redescriptions that make evident what should be obvious to anyone not led astray by philosophical theorizing. They offer no new information; rather they are, '...an attempt to draw our attention to something that lies before our eyes'.[15]

Wittgenstein's early treatment of logic provides the best and clearest example of how philosophy can clarify matters and expose philosophical nonsense, without advancing any kind of theory. To fully appreciate this it is useful to consider the context in which this approach developed. He writes:

> All philosophy is a 'critique of language'....It was Russell who performed the service of showing that the apparent logical form of a proposition need not be its real one. (TLP 4.0031)

In praising Russell's good work, Wittgenstein is, of course, applauding the kind of analysis that lies at the heart of the theory of descriptions. On that approach, troublesome definite descriptions – such as those that seemingly pick out non-existent entities or impossible objects – are logically analysed to reveal that they are in fact non-referring. Such items generated terrible problems for the metaphysics of Platonic Atomism, which all too easily became committed to the existence (or more precisely subsistence) of impossible objects simply by mentioning them in coherent propositions. For example, it seems that we need to posit the existence of round-squares in order to account for the fact that we can sensibly claim that 'The round-square cannot be'. Famously, in Russell's later writings such propositions were analysed as asserting something quite different. Informally, the content of (1) could be represented by (1').

(1) 'The round-square cannot be'

(1') 'There is no such unique entity x which has the properties of being both round and square'.[16]

In performing this kind of logical analysis Russell was able to explain the meaningfulness of propositions like (1) without having to make ontological space for impossible objects. The apparent name 'round-square' was thus treated as an incomplete, non-referring symbol. On its own it does not name anything. However, when we understand its use properly we can see what is being asserted by the entire proposition. By considering cases such as these we can see why Russell drew a distinction between what we seemingly say (grammar) and what we really assert (logic). We can see then a definite evolution in Russell's views on these matters from his Platonic Atomist days, when he held that terms were the building blocks of reality. During that period he held that we were directly acquainted with the constituent terms of any given proposition and that language was a transparent medium (a clear window, as it were) through which we 'perceived' this intellectual reality.

Despite these developments in his understanding of language, Russell continued to postulate the existence of logical forms, which he thought were needed in order to account for the relations between objects and, indeed, our capacity to make judgements about those relations. Thus he maintained that the propositions of logic had a kind of Platonic status, as independent forms that could be described separately from their contents. According to Russell's logical atomism, we must treat only the fully analysed entities and the logical components of genuine statements as having real, metaphysical status. In his abandoned 1913 manuscript, *Theory of Knowledge,* he had begun to develop the view that we must be acquainted with logical forms, such as aRb, in making various kinds of judgements. It was his view that discovering and charting these possible forms of judgement was a positive and important task for philosophy.

The novelty and ingenuity of Wittgenstein's early philosophy was to take this account a step further and to challenge what we might call, following Peterson, the naive representationalist view of logic. Indeed, that there are no logical objects is his self-avowed fundamental thought (*Grundgedanke*).[17] Thus, his analysis of the so-called propositions of logic reveals them to be, strictly speaking, without sense. If we crudely characterize Russell as a Platonist, then by contrast, we can regard Wittgenstein as adopting a more Aristotelian line.[18] For him, logical form is immanent in the structure of our ordinary statements, it is not something 'extra' that needs to be added or otherwise attached. To use an ordinary analogy, it is akin to the *way* in which ingredients are blended together to make a cake. One does not add the eggs, flour, sugar and so on and then add in 'the mixing'.

In the prelude to his attack on logical objects Wittgenstein introduced the idea of formal concepts, which are to be contrasted with proper concepts. He writes:

> When something falls under a formal concept as one of its objects this cannot be expressed by means of proposition. (TLP 4.126)

The thought is that the nature of a formal concept can only be shown in the way that it is *employed*. It cannot be said (which, if we take his other remarks about saying seriously, would involving picturing a possible state of affairs). Hence, he writes: 'Thus the variable name 'x' is the proper sign for the pseudo-concept object' (TLP 4.1272).

What Wittgenstein is telling us is that there is no 'thing' for which the apparent name 'object' (or variable, x) is a proxy. There is no super-object that answers to the name x. In this way, our understanding of the general term 'object' can only be shown in the use we make of the logical variable when it is employed in relation to genuine objects that can be substituted for it. The components of the so-called propositions of logic are to be regarded as purely formal in this sense. Logical 'names' have no corresponding objects and logical 'propositions' picture no states of affairs. Hence his fundamental thought amounts to the claim that the seeming propositions of logic are, in fact, non-representational.[19]

This is all quite familiar, but what matters to our discussion is the way in which Wittgenstein attempts to free our thinking on these matters. He does not simply advance an argument or premises, nor does he put forward a theory of logic. Rather, in large part, he presents us with an alternative notation that is designed to show us that certain kinds of logical symbolism can obscure the fact that logic has a purely formal character. For example, by using symbols for logical constants, Russellian notation can mislead one into endorsing a mythological ontology, especially if one has other explanatory commitments.

The truth tables are meant to break the spell cast by Russell's symbolism. For instance, 'if p then q' can be represented by using the truth table as a structurally adequate formulation in which the logical constant for the conditional does not appear at all. Consequently, when presented with truth tables, we are forced to focus on the *use* of logical symbols. This alone should cure us from mistakenly thinking of them in terms of their capacity to represent. We are told: '...in fact all the propositions of logic say the same thing, to wit nothing' (TLP 5.43).

By approaching the problem in this way, Wittgenstein holds out the possibility of a cure for the practice of mythologizing logical objects. At the very least, he removes at least one major temptation for thinking that there must be logical objects represented by the logical constants.[20] In playing this role the truth tables provide paradigm examples of a formal, perspicuous elucidation. They tell us nothing new; they merely, and quite literally, re-present that which already lies open to view. They are well-formed elucidations that say nothing. Yet, they get us to see things differently by showing us other possibilities. They make evident something that we should be prepared to recognise but would, otherwise be unable to see.

Further support for this reading is provided by Peter Hylton who provides a careful analysis of the potentially confusing passages of the TLP 5.2s, and 5.25 in particular, in which Wittgenstein warns us not to confuse operations and functions. To make sense of this otherwise strange contrast, Hylton convincingly argues that it was drawn in order to distinguish Wittgenstein's views on the nature of truth functions and operations from those of Russell, and only incidentally those of Frege.[21] This is because, specifically in line with his views on the independent status of logical forms, Russell held that propositional functions were used to generate more complex representations from simple elements. This is not surprising if we take the view that logic is representational. For example, on such an approach, '...the propositional function X is wise...share[s] a structure with the proposition that Socrates is wise: the propositional function is not a mere mapping of objects onto propositions'.[22] This, of course, was anathema to Wittgenstein and it is precisely what his n-operator for truth functions is designed to avoid, by demonstrating that logical operations are purely formal. It is introduced, in part, to show how complex propositions can be formed from elementary propositions without introducing or requiring any extra or more complex representational resources. Thus it is not surprising that, immediately following his remarks on logical operations, he writes: 'At this point it becomes manifest that there are no 'logical objects' (in Frege's and Russell's sense)' (TLP 5.4).

But, if the logical sections of the *Tractatus* provide the best examples of clarification at work how do they compare with the other sections of the book? Consider what is heralded in its opening line and early remarks:

> The world is the totality of facts, not of things. (TLP 1.1)
> What is the case – a fact – is the existence of states of affairs.
> A state of affairs (a state of things) is a combination of objects (things).
> It is essential to things that they should be possible constituents of states of affairs. (TLP 2–2.011)

In order to understand the character of these remarks, it is important to consider the context from which they emerge. If Hylton is correct in supposing that it was with Russell in mind that Wittgenstein presented his account of the n-operator, then it is also likely that these early remarks are meant as a critique of Russell's ontology. In his *Lectures on Logical Atomism*, Russell claimed that we could, '...get down in theory, if not in practice, to ultimate simples'.[23] These 'simples' were regarded as the basic constituents of the world – its logical atoms. Moreover, they were the only objects that could be given true names. Of course, Russell realized that this was neither a satisfactory nor complete inventory of the furniture of the world. He recognized that, '...the only other sort of object you come across in the world is what we call facts'.[24] Yet he qualified this by saying that facts, '...are not properly entities at all in the same sense in which their constituents are. That is shown by the fact that you cannot name them'.[25] Nevertheless, according to Russell, the upshot is that our basic ontology must be one of objects and facts.

Against this background, the *Tractatus* appears to challenge Russell's claim by suggesting that it does not follow that objects comprise an independent ontological category, even if we accept that facts are composed of objects. For we are quickly told that it would be impossible to imagine any 'thing' outside some possible situation (that is, outside a given factual context). Consequently, objects are not independent of facts, but nor are facts independent of objects. Objects are always in one situation or another and, in being so, they constitute facts – which are nothing other than the way various things stand in relation to one another.

Although the *Tractatus* is undoubtedly critical of Russell's views in this respect, there are two importantly different ways of thinking about the nature of this criticism. On the one hand, Wittgenstein might be seen as offering a straight theoretical or metaphysical adjustment to Russell's position in the hope of simply 'building a better mousetrap'. Thus he may be thought to be attempting to improve on the theories that Russell and Frege had already established. Read in this way, the opening remarks, and those that constitute the so-called picture theory, are primarily correctives offered with the same philosophical spirit and end in mind.

More radically, these remarks can be taken as a wholesale rejection of the entire project of philosophical theorizing. That is, Wittgenstein can be seen as rejecting Russell's approach to, and vision of, philosophy. On an anti-theoretical reading, although the same criticisms are being made, they are not advanced as part of a developing theory with its own set of new and improved posits. Instead of replacing Russell's 'things and facts' with his 'facts', Wittgenstein could have been trying to discourage this entire style of approach. If he was doing this, then he was both trying to get us to see things correctly and critically responding to Russell's type of metaphysical posturing. Marie McGinn gives an excellent account of the status of these remarks, in defending the idea that there is an elucidatory core of the work, which is not composed of mere attempted statements of fact.[26] She writes:

> The principal application he makes of the concrete image of a world of facts which consists of objects in combination with one another is to use it as a means to make clear the distinction between content (objects), structure (the arrangement of objects in determinate relationships to one another in facts) and form (the possibility of objects entering into these determinate relationships).[27]

Although Wittgenstein necessarily uses language to clarify these points, his sentences, '...are not putting forward a theory of the proposition or a speculative account of the relationship between language and the world'.[28] Read in this light, even TLP 1.1 sets its face against philosophical theorizing as such.

In line with his understanding of logic, this would also account for Wittgenstein's silence on the question of how things must be combined in states of affairs and what kind of things there are. As Lynette Reid writes:

> One of the anti-metaphysical strands of the *Tractatus* is that logic cannot judge in advance what the internal articulation of fully analyzed propositions will be: contrary to Frege and Russell, who think it essential to the nature of representation that a proposition segment into subject and predicate of some sort, the *Tractatus* denies that there is any point in discussing in advance whether elementary propositions will consist of names and concept-expressions, or n-termed relation-expressions, or anything else.[29]

This would explain why Wittgenstein provided no examples of 'objects' or 'elementary propositions'. Although this has long puzzled commentators who adopt a theoretical reading, in accepting that his purpose was not the scientific one of providing a metaphysical inventory of reality or of articulating the true logical structure of the world, it becomes clear why he did not provide any such examples.[30]

The Revocation

There is an apparent tension in the *Tractatus*, which is notoriously identified by Wittgenstein at its very close. He writes:

> My propositions serve as elucidations in the following way: anyone who understands me eventually recognises them as nonsensical, when he has used them – as steps – to climb up beyond them. (He must, so to speak, throw away the ladder after he has climbed up it). (TLP 6.54)

On the doctrinal reading these remarks suggest that, as a work of philosophy, the *Tractatus* attempts to say what cannot be said. It appears to be a series of intelligible propositions that, unlike genuine propositions, employ pseudo-concepts, such as object, fact and so on.[31] Yet being so, by its own account, it must be condemned as nonsense. For the book itself tells us that all propositions serve to picture facts and must have the potential to be true or false, depending on whether or not things happen to be as they say. This condition must hold in order for a proposition to have a sense. As this is the general form of propositions, then no proposition can say anything necessarily true. Yet, if we understand (and accept) what its propositions seemingly say – especially those concerning the way in which propositions picture reality – then we will realize that the entire book is an attempt to say what cannot be said. Its propositions are improper if we regard them as unargued theoretical pronouncements. They are, strictly speaking, nonsensical because we cannot imagine a possible (Tractarian) world in which they do not hold true. They seemingly tell us things like: the substance of the world resides in simple, indestructible objects; these objects combine to create states of affairs; the states of affairs that actually hold are the facts of the world and so on. Put directly, if we treat such 'propositions' as meaningful then we must at the same time reject them as meaningless.

However, armed with the idea that the apparent philosophical statements are supposed to be merely clarifactory elucidations, things look different. Despite their outward form, those elucidations are merely pointers designed to remind us of what we ought already to know. In this respect, Wittgenstein specifically contrasts them with the kind of factual statements that are meant to typify the corpus of the natural sciences. As he sees it, it is the business of science to tell us exactly which states of affairs contingently obtain in the world. In this respect, reality is its exclusive concern. Consequently:

> The totality of true propositions is the whole of natural science (or the whole corpus of the natural sciences). (TLP 4.11)

But, in what might otherwise appear to be an attitude of deference to science on matters of importance, Wittgenstein denies that we ought to be interested in reality *per se*. It is possibility, not reality that is his concern. Logic alone, which says nothing, defines the limits of possibility and also the limits of thought. Indeed, even as late as 1930, we find him endorsing this view. He writes:

> It is all one to me whether or not the typical western scientist understands or appreciates my work, since he will not in any case understand the spirit in which I write. Our civilisation is characterised by the word 'progress'. Progress is its form rather than making progress being one of its features. Typically it constructs. It is occupied with building an ever more complicated structure. I am not interested in constructing a building, so much as having a perspicuous view of the foundations of possible buildings. So I am not aiming at the same target as the scientists and my way of thinking is different from theirs.[32]

The contrast between logic and science and their respective spheres helps illuminate the difference between philosophical elucidations and ordinary propositions. Yet, most of his elucidations are ill-formed given that they should, like logic, say nothing. It repays us to attend to the remark that proceeds the revocation, where he says:

> The correct method in philosophy would really be the following: to say nothing except what can be said, i.e. the propositions of natural science – i.e. something that has nothing to do with philosophy.... (TLP 6.53)

Yet, even if we adopt this anti-metaphysical elucidatory reading, the paradox remains. Whereas it can shown that there are no genuine propositions of logic, the elucidations concerning the essential nature of propositions, thought and language are meant to be necessarily true, yet contentful observations. If Wittgenstein's elucidations concerning sense and content hold fast, then no contentful proposition could serve as a philosophical elucidation. To serve such a purpose they would need to say the unsayable. Consequently, on either reading, what is written in the *Tractatus* is in violation of its own account of sense.

In this light, it may appear to be a virtue of the purely therapeutic reading that it resolves this internal paradox. Both Conant and Diamond challenge the idea that Wittgenstein seriously wished us to recognize something which is 'nonsensical but significant', even in his early period.[33] For them, from beginning to end, Wittgenstein regarded all nonsense as 'garden-variety' gibberish.[34] Thus they maintain not only that Wittgenstein's aim to do away with philosophical nonsense remained constant throughout his philosophical career, but, more strongly, that his understanding of nonsense remained constant as well.

On their approach, we have the means to make sense of Wittgenstein's plea that we 'must throw away the ladder' without committing ourselves to the idea that the work culminated in an irreconcilable contradiction. We are asked to focus on the fact that he self-consciously employs a 'strictly incorrect' method for therapeutic purposes.[35] The recognition of the paradox is yet another means of getting his readers to see the impossibility of traditional philosophy. First, it lets them imagine they can find sense in its pseudo-propositions and then it pulls away the rug by showing that endorsing such a view leads to no good end. To effect the cure properly requires giving the patient enough rope.[36] On this view, the self-destructive character of the *Tractatus* is intended as part of the therapy. Thus, Diamond sees the key to unlocking Wittgenstein's remarks of TLP 6.54 in the sentence, '...anyone who understands me eventually recognises them as nonsensical'. She makes great play of the fact that he writes about understanding *him* as opposed to the understanding the propositions of the *Tractatus* itself.[37] Furthermore, she claims that the remarks in the preface and in the last section act as part of the book's frame: they provide instructions for reading the rest. Those at the beginning tell us that the project is to set a limit to what can be said wholly from inside the bounds of sensible language. Similarly, the closing remarks are a final reminder of this message – for those who might be tempted, despite his rare asides, to misunderstand his true project and purpose.

But this reading is only superior if we think it best to resolve the tension. I am not convinced that it is. There are good reasons to think that the *Tractatus* is best seen as an imperfect attempt to offer elucidations precisely because some of them are both ill-formed and infected with faulty assumptions about the nature of language and the limits of sense. The major assumption that needs to be rejected is that all propositions serve to represent facts and that all sensible language shares this uniform function. Indeed, this view underpins the bulk of the book and infects the remarks in sections other than those concerning logic, such that they are failed attempts at clarification.

Indeed, the scene for the revocation only makes sense if we understand it against the backdrop of Wittgenstein's uniform and restrictive vision of the function of propositions, for it was this that gave definition to his early accounts of sense and nonsense. There is no such tension in the later writings because his views about sense became more liberal as they developed. On the standard account, he went on to reject the family of views associated with logical atomism, including the ideas that propositions have a general form and that they are determinately, and independently, true or false. Instead of these views, he came to favour the idea that to understand the functions of language we must look to its contexts of use. Indeed, it is only by recognizing this that we can gain insight into the important similarities and differences that exist between his early and mature philosophy.

On this important point I agree with doctrinalists. For it would be difficult to deny that his views on the nature of sense did so evolve. How else can we interpret what he tells us in the preface to the *Investigations* when he remarks that, '...since beginning to occupy myself with philosophy again, sixteen years ago, I have been forced to recognize grave mistakes in what I wrote in that first book'?[38] If the book was not trying to 'say' anything at all then what mistakes could be present in what he had *written* in it?[39] Moreover, there has been important scholarship that appears to successfully identify the cause of this shift in his views. For example, Jacquette traces the main catalyst for this to Wittgenstein's inability to deal adequately with the colour incompatibility problem, which Ramsey brought to his attention.[40] My point is that if we were to adopt an overly strong therapeutic reading we would be unable to make sense of this transitional phase for, on such a reading, there should be no such transition.[41]

Furthermore, accepting that such a change took place has other potential explanatory virtues. For example, it may help to explain the dramatic changes in Wittgenstein's writing style. From what we know of him, such changes would not be philosophically trivial. As McDonough notes, 'Wittgenstein regarded himself as an artist as well as a philosopher. [He] was obsessed with stylistic questions, as well as questions about the origin of his own style'.[42] Thus the fact that in the *Tractatus* his remarks have the outward form of tightly compressed propositions in a numbered sequence is hardly surprising if he took seriously the idea that this was the only true form for the expression of thoughts. Nevertheless, given his therapeutic aim, he was self-conscious that his style was 'strictly' incorrect. In his later work, as his view of thought and language became more open, his style, too, became more natural and free. We find therein a host of aphorisms and comparisons in the form of ordinary sentences, questions, paragraphs and short passages. Numbered propositions have disappeared.[43]

Given this, we should read his remark at TLP 6.54 not as a note of guidance, but as a recognition of a genuine tension. For, although he had a firm view about how philosophical problems had to be solved, the fact was that the *Tractatus* failed to provide the basis for completely satisfactory solutions. That Wittgenstein recognized his method to be 'strictly incorrect' also fits well with the fact that he expresses himself by asking his readers to understand him or his intention, rather than what is written in the book. In accepting that the work was meant to be therapeutic, we need not hold that he made no erroneous assumptions or that he was successful in prosecuting his intended end. Although he wished the *Tractatus* to be a work of clarification, in large part, it failed this purpose. This is important, for if we view the revocation as highlighting a tension that is genuinely present in his work we can regard it as the seed for the future development and refinement of his philosophical views and style.[44]

In this light, even though there is something right about the therapeutic reading it should not be fully endorsed. Rather, what is required is a reading according to which it is possible to acknowledge the continuity of Wittgenstein's clarificatory aim and method, while at the same time allowing that his thought developed and matured. I believe that such a reading is available to us if we focus on his replacement of logical form with forms of life as the governors of sense.

From Logical Form to Forms of Life

In my view, the ways in which Wittgenstein thought philosophy, logic and ethics and aesthetics 'say nothing' were importantly different. We have already considered the cases of logic and philosophy but, before considering ethics and aesthetics, it is important to mark the changes in his views concerning sense and nonsense. The crucial change was the replacement of logical form as the transcendent basis of language by forms of life. Famously, in the *Investigations*, in contrast to the idea that all propositions have a common essence underpinned by their logical form, we are told that, '...to imagine a language means to imagine a form of life'.[45] Elsewhere we are instructed that 'Here the term "language-game" is meant to bring into prominence the fact that the speaking of a language is part of an activity, or a form of life'.[46]

My claim, in this section, is that, despite some important differences, forms of life play essentially the same role in the later writings that logical form played in the *Tractatus*. Crucially, both govern 'the bounds of sense' and must remain outside the scope of the explicable, strictly sayable or articulable. Neither logic nor grammar can make any pronouncements, as they are the transcendental limits to sense. For this reason, these limits cannot be stated or positively charted. Evidence of this similarity of role is found in what Wittgenstein says about these two notions.[47] For example, consider his ill-labelled 'doctrine of showing', as it makes an early appearance in the TLP 4s. We are told that:

> Propositions can represent the whole of reality, but they cannot represent what they must have in common with reality in order to be able to represent it – logical form.
> What expresses itself in language we cannot express by means of language.
> What can be shown cannot be said. (TLP 4.12–4.1211)

By insisting that what can be shown cannot be said, Wittgenstein was preparing us for the idea that we cannot say anything about the logical form of propositions. Yet, as we can see by Russell's remarks, the standard opinion was that nothing in principle should prevent this. All that is required is the construction of an appropriate meta-language. Hence, Russell writes:

> In my introduction to the *Tractatus*, I suggested that, although in any given language there are things, which that language cannot express, it is yet always possible to construct a language of a higher order in which these things can be said.[48]

As we have already seen, the Russell–Wittgenstein debate about saying and showing is not an idle one: it is yet another expression of their different attitudes concerning the status of logical propositions. For Wittgenstein, logic had no objects and no content, it could not inform us about the nature of the world. The apparent statement 'P iff Q' is not a claim about how things stand; rather, it specifies how two facts must be related or structured.[49] But to accept this is to put to the sword the very idea that there could be any higher-order language that could express that which is unsayable. In suggesting otherwise Russell failed to fully grasp the radical character of the Tractarian position and its view of logic. Nevertheless, some remarks in the *Tractatus* concerning the transcendental priority of logic appear to go against this reading. Thus, we are told: 'Logic is not a body of doctrine, but a mirror image of the world. Logic is transcendental' (TLP 6.13). However, although Wittgenstein regarded logic as transcendental, he did not see it in the Russellian sense of being separate but rather in the sense that it was the limit of sensible language; the stopping point. Consider that he also says:

J's target

log. & gr. transcendental

> Logic is prior to every experience – that something is so. It is prior to the question 'How?', not prior to the question 'What?' (TLP 5.552)

This is a reminder that although logic is transcendental, in that it constitutes the limits of what is possible, thinkable and sayable, it says nothing about the nature of what is found in the world. Logic is at one and the same time built into the nature of things and the nature of language; it is their common link. Exactly how things can be arranged, in fact, is determined by the nature of the things themselves. This is their logic. Therefore it could not be something additional.

Wittgenstein is certainly not saying that the scaffolding of the world is separable from it or that we can describe it independently. Hence, 'Logic pervades the world: the limits of the world are also its limits' (TLP 5.61). Logic cannot exist independently, for it has a merely structural character. It follows from this that there could be no distinct 'science of logic', as logic is not something that can be articulated. Nor could it be a source of explanatory laws of inference, for we are told that:

> Our fundamental principle is that whenever a question can be decided by logic at all it must be possible to decide it without more ado. (TLP 5.551)
> Logic must look after itself. (TLP 5.473)

We must accept this if we are to avoid an infinite regress. For, if the propositions of logic did say something about the world then they too would have to be underwritten by some kind of superior logic. If the propositions of such a superior logic were to say something then they would also need to be underwritten by a yet more superior logic, and so on. In this way, if logic underwrites all saying it must not say anything itself. For this reason we are told that: 'Clearly the laws of logic cannot in their turn be subject to the laws of logic' (TLP 6.123).

For our immediate purposes, what is interesting about these remarks concerning logical form is the parallel that can be found between them and those that he makes about language games, grammar and forms of life. As with logic, we are advised to '[l]ook on the language game as the *primary* thing'.[50] We are also told: 'What has to be accepted, the given, is – so one could say – forms of life.'[51]

Furthermore, we find a parallel to the idea that logic cannot provide explanations of inference echoed in his later treatment of rule-following. There he bids us to consider the positive role that ostensive teaching plays in enabling us to set standards in order to avoid an infinite regress. To make this case fully would require a great deal more space, but a few suggestive comparisons should suffice to make the initial point. Consider that, at base, rule-followers do not rest their interpretations on other interpretations. For this reason, learning how to use concepts does not require any knowledge in the initial stages. Instead, it presupposes that there are certain capacities and abilities that are common to both the teacher and student and that the latter is willing to blindly obey authority. We are told that:

> Following a rule is analogous to obeying an order. We are trained to do so; we react to an order in a particular way....[52]

Crucially, the regress is blocked because learning to follow a rule does not presuppose the kind of intellectual capacities that such training is meant to engender. The novice becomes an autonomous speaker by learning a skilled technique within a social medium. Furthermore, the problem of multiple interpretations is also tamed for, in the process of such training, a shared sense of the *obvious* is developed, which is based on, but not confined to, our common natural reactions.[53] Given this, we develop a communal second nature via training, as Aristotle once proposed.[54] It is this social stage-setting that makes meaning and rule-following possible. If we fail to understand this then we will be plagued by unworkable and misleading philosophical pictures that fail to satisfy.[55] Hence, just as there can be no laws of logic that can explain or externally justify our inferences, so there can be no rules of language use that can provide explanations or justifications of our linguistic practices. Wittgenstein anticipates our tendency to go wrong just here when he writes:

> Our mistake is to look for an explanation where we ought to look at what happens as a 'proto-phenomenon'. That is, where we ought to have said: *this language-game is played*.[56]

In this light, we may wonder what has really changed. It appears that the real difference between the early and late periods is that instead of objects fixing logical possibilities, and in turn determining what can be sensibly said, in the later writings it is grammar that performs this function. Thus:

> Grammar tells us what kind of object anything is.
>
> Essence is expressed by grammar.[57]

There are still more similarities. For just as there was no way of charting the limits of logic independently, there is no point in trying to understand forms of life, language games or grammar from on high. There is no getting behind, beneath or above grammar for the philosophical purpose of providing some kind of overview. Instead, we are reminded that:

> The limit of language is shown by its being impossible to describe a fact which corresponds to (is the translation of) a sentence without simply repeating the sentence (This has to do with the Kantian solution to the problem of philosophy).[58]

Even more importantly, these similarities are not accidental. They can be explained if we consider that in both periods he was steadfastly engaged in grammatical investigations, always with the aim of clarifying and getting a clear view.[59] His aim was ever to prevent the bewitchment of our intelligence by language.[60] Nor is this surprising if the 'fundamental thought' of the *Tractatus* was still driving the later Wittgenstein. Considered thus, the move from logical form to forms of life was a natural, almost irresistible consequence of his growing interest in the distorting effects not just of logical constants, but also of other forms of symbols, such as ordinary words and names. We can regard the approach in the later writings as a wider application of the proper method for treating philosophical problems, as exemplified by the elucidations concerning logical symbolism. Read in this way, the development of this early approach paved the way for the kind of case by case analyses that populate the later writings.

In abandoning the idea that propositions have a general form, it also becomes necessary to abandon the monolithic conception of logical form that supposedly underpinned them. For example, something dramatic happens to the early account when we remove the view that propositions are essentially representative – that is, a certain picture of logic loses its place as the ground for sense. Hence, unsurprisingly, in the *Investigations* he writes:

> But what becomes of logic now? Its rigour seems to be giving way here. – But in that case doesn't logic altogether disappear?...The philosophy of logic speaks of sentences and words in exactly the sense we speak of them in ordinary life when we say e.g. 'Here is a Chinese sentence', or 'No, that only looks like writing; it is actually just an ornament' and so on. We are talking about the spatial and temporal phenomenon of language, not about some non-spatial, non-temporal phantasm.[61]

With logic no longer playing a foundational role we need to look elsewhere to find the crucial link between language and the world, and what governs sense. That is precisely why forms of life come to replace logical form as the ground for connecting language and the world. Moreover, once freed of his one-dimensional vision of the function of language, Wittgenstein was forced to give careful attention to the operation of other forms of language as well. Although these are important changes, they are often misrepresented in such a way as to cause us to overlook the continuity in his overall method. We still find him trying to overcome the traps set by the misleading character of symbols by focusing on the use of language. Consider his early remarks in the *Philosophical Investigations* in which he famously writes:

> It will prove useful in philosophy to say to ourselves: naming something is like attaching a label to a thing.[62]

His treatment of names echoes his treatment of the logical constants. For, just as logical symbols mesmerize us, so can ordinary names. He wants to make us aware of how much we presuppose when we employ the picture that language is a system of signs that stand for, or represent, objects. He reminds us that:

> Naming is so far not a move in the language-game – any more than putting a piece in its place on the board is a move in chess. We may say: *nothing* has so far been done, when a thing has been named.[63]

We are urged to think of words not as mere labels but akin to the handles and knobs in locomotives. These all have many different jobs that must be understood in relation to one another.

Furthermore, attention to the early work sheds light on his use of alternative language games as a philosophical tool.[64] In such cases, we are asked to imagine situations in which 'others' use concepts differently than we do for the express purpose of getting a clear view of the grammar of our own language.[65] Consider, for example, why he introduces us to the possibility of other language games that might surround the concept pain, as in the extreme case described in *Zettel*[66] where he conjures up a tribe which employ two different concepts of pain: '...one is applied where there is visible damage and is linked with tending, pity and so on. The other is used for stomach-ache for example, and is tied up with mockery of anyone who complains'.[67] Accordingly, unless members of this tribe can locate some kind of outer bodily damage they will not regard the person as experiencing what-we-would-call-'pain'. Their notion of what counts as pain cuts much more finely than would ours. Or, as Wittgenstein puts it, they, '...have concepts which cut across ours'.[68] I claim that at least one crucial aim of these exercises in imagination is to get us to realize that 'We are not analysing a phenomena (e.g. thought) but a concept (e.g. that of thinking), therefore the use of the word'.[69] For example, this helps to break us of our natural temptation, if we hold that language serves to name objects, to try to observe the processes that go on when we are thinking. That is, we try to observe some *thing* that the word *thinking* names. Thus, he notes:

> In order to get clear about he meaning of the word 'think' we watch ourselves while we think; what we observe will be what the word means. – But this concept is not used like that.[70]

He tells us: 'What we deny is that the picture of the inner process gives us the correct idea of the use of the word "to remember"'.[71] But in denying that there is some *process* or *thing* that the term 'thinking' designates, Wittgenstein is emphasizing what should be of interest to us when we wish to investigate the nature of our psychology. He writes:

> [W]e forget that what should interest us is the question: how do we *compare* these experiences; what criterion of identity *do we fix* for their occurrence?[72]

We are reminded that 'We learn the word "think" *under particular circumstances*'[73]and also that 'The surroundings give it its importance'.[74] It is precisely for this reason that our use of language is not *arbitrary*, relative or merely conventional. This is why 'One cannot guess how a word functions. One has to *look at* its use and learn from that' and why he urges us to, 'Let the use of words teach you their meaning!'.[75] This use of language games is reminiscent of his use of truth tables in the *Tractatus*. In this light the later writings are natural evolutions from his early work.

This is not to deny that there are important differences in the early and late periods. Indeed, these also need to be emphasized. But this is only possible by giving detailed attention to the effects of the move from logical form to forms of life.

Usefully, in considering to what extent Wittgenstein ought to be thought of as a transcendental philosopher, Meredith Williams compares and contrasts his philosophical approach with that of Kant. The similarities are clear. Both are concerned with issues concerning the bounds of sense such that it makes it look as if Wittgenstein's grammar could be a substitute for Kantian *synthetic a priori* categories. But the key difference between them is that, even in his early philosophy, logical form, which says nothing, could not be equated with Kantian categories, if these are regarded as defining a positive limit to the bounds of sense.

Even more so, the later Wittgenstein had an obviously fluid vision of what drives conceptual change and what underlies the boundaries of sense. Accordingly, in his famous discussion of the concept of 'game', we are told that, '...the extension of the concept is not closed by a frontier....For how is the concept of a game bounded?...Can you give the boundary?'.[76] Thus:

> We do not know the boundaries because none have been drawn. To repeat, we can draw a boundary for a special purpose. Does it take that to make the concept usable? Not at all.[77]

Concepts get their lives from our practices, not vice versa. This is why it is mistaken to treat 'communities' or 'grammar' as fixed. Since these practices develop and evolve, there can be no transcendental setting of limits to sense in advance or once and for all. It is for this reason that philosophers cannot theorize from the general to the particular, but must instead merely describe and be vigilant of transgressions of sense. In this he, '...inverts the Kantian order of priority'.[78]

Bearing these points in mind, we are now better armed to consider the nature of his remarks on ethics at the end of the *Tractatus*.

Remarks on Value

We can throw light on Wittgenstein's early remarks concerning ethics, aesthetics and religion by focusing on his changed views about the nature of sense. This is vital, since these remarks are regularly identified as being the very heart of the work.[79] As with logic, we are told that these domains are both transcendental and nonsensical. But, as Peterson notes, they are non-representational in a different way than logical 'statements'. He rightly insists that they involve 'what lies outside of language' whereas logic concerns what is internal to it (given the Tractarian view).[80]

In advancing her strong therapeutic reading, Diamond claims that Wittgenstein's understanding of nonsense in both periods is uniform, and effectively equates to the 'technical' sense of nonsense, found in the Tractatus. Therefore given that his remarks on ethics are designated as nonsensical they must be treated just as any other kind of nonsense. But, as we noted earlier, for Diamond there is only one kind of nonsense.

One paradox that emerges in advancing this view is that in order to take seriously Wittgenstein's early views on nonsense it seems we must also take seriously his remarks on propositions serving as pictures of facts. But this would require us to read those sections of the Tractatus non-ironically, against the counsel of the therapeutic reading. Even more seriously, the strong therapeutic reading makes it difficult to see why he places such importance on the final sections of the work if they are simply nonsensical remarks to be overcome and thrown away. Diamond, to her credit, attempts to circumvent this problem by distinguishing the attractiveness of speaking ethical nonsense as opposed to speaking other forms of nonsense. She treats, '...cases of understanding a person as saying in his heart something that makes no sense, [as] something which we have the imaginative resources to grasp as attractive where that imaginative capacity is tied to our own capacities as moral agents'.[81] She writes:

> ...if we read the *Tractatus* right, the upshot of the book will be different in regard to the two sorts of utterers of nonsense. The attractiveness of philosophical sentences will disappear through the kind of self-understanding that the book aims to lead to in philosophers; the attractiveness of ethical sentences will not. But if we understand ourselves, ourselves the utterers of ethical nonsense, we shall not come out with ethical sentences under the illusion that we are talking sense.[82]

But if all speaking nonsense is really on a par – that is, it is all like saying 'piggly-wiggly' – then how can uttering one bit of nonsense be more attractive than uttering another?

Diamond is right in that to think that Wittgenstein was suggesting that there could not be ethical 'doctrines' or 'propositions'. However, what we should not lose sight of is that her reading also rules out a more sophisticated understanding of the ethical remarks – one which makes sense of Wittgenstein's evolving conception of language. For example, the fact that he regarded ethical remarks as strictly nonsensical but nonetheless important is best seen as a vital insight into the essence and function of ethical language, albeit a negative one. That is, Wittgenstein realized that they, too, are non-representational in that they do not serve to represent facts.[83] Yet, he saw that the silence of ethics was pregnant in a way in which the silence of logic was not. That is, he struggled to maintain that ethics had a profound status, despite being nonsensical in Tractarian terms. His emphasis was on its profundity. This was tied to the fact that ethics is nonsensical only to the extent that he recognized that it did not involve factual statements. The problem was that, given his uniform account of sense and nonsense, he did not have the resources in his early philosophy to properly develop his positive view of ethics. This was a source of genuine tension for him. Yet, as Wittgenstein's views on the essence of language evolved this and other important arenas of discourse no longer had to be regarded as nonsensical merely because they lay outside the bounds of the factual. He came to recognize that boundaries may be drawn for, '...various kinds of reason'.[84] Once freed of the Tractarian vision of the function of language, the later Wittgenstein was better able to explicate his views concerning the nature of nonsense itself. On this reading he was aware from the outset, even if only negatively, that ethical talk is not a form of 'factual discourse'. This fits well with his more mature position concerning language games that relate to particular domains and the attention he demands that we give to the context of our activities. If we focus on the changes in his account of sense as he moved from logical forms to forms of life, we can see how he advanced from a mere recognition that ethical language was non-representational and began to focus more positively on the contexts in which such remarks occur.

However, seeing these developments in his thought is ruled out by a reading that doesn't allow us to take his views on the general form of propositions seriously. When we consider the links *and* breaks between the early and later writings, it becomes clear that it is a mistake to think that all that is regarded as nonsensical in the *Tractatus* can be treated alike.[86] As Reid puts it, in this light we can see, '...a strong warning from the perspective of his later thought that the frame of the *Tractatus* is not the expression of a clear grasp on the part of the author of the *Tractatus* of what it is to label some use of language "nonsense"...'.[85] Moreover, the very fact that Wittgenstein was prepared to allow for the existence of profound forms of nonsense in his early work, despite being unable to incorporate this idea seamlessly into his thinking, is vital to an understanding of the man, what he found important and how his thinking progressed.

Notes

* I would like to thank Jo Wolff for allowing me to reuse and expand upon material from my section of the paper, 'Making Sense of Nonsense: Kierkegaard and Wittgenstein', written jointly with John Lippitt, and published in *Proceedings of the Aristotelian Society*, **XCVIII** (3).

1 Ludwig Wittgenstein, *Culture and Value*, ed. G.H. von Wright and H. Nyman, trans. P. Winch (Oxford: Blackwell, 1980), p. 20e.

2 An example of this kind of view is inherent in Marie McGinn's claim that these works are '...both clearly by the same author, but one whose *conception of his philosophical task* undergoes a profound change' (Marie McGinn, *Wittgenstein and the Philosophical Investigations*, London: Routledge, 1997, p. 2, emphasis mine). Or similarly, according to Hacker, '...what is much more important is that the *Investigations* as a whole stands opposed to the philosophical *spirit* of the *Tractatus*' (P.M.S. Hacker, *Wittgenstein's Place in Twentieth-Century Analytic Philosophy*, Oxford: Blackwell, 1996, p. 98, emphasis mine).

3 James Conant, 'Must We Show What We Cannot Say?', in R. Fleming and M. Payne (eds), *The Senses of Stanley Cavell*, Bucknell Review (Lewisburg, PA: Bucknell University Press, 1989), pp. 248, 266.

4 Cora Diamond, *The Realistic Spirit: Wittgenstein, Philosophy and the Mind* (Cambridge, MA: MIT Press, 1991), p. 18.

5 Ibid., pp. 181–82; cf. also p. 194. For this reason she is openly hostile to standard readings because, '...the attempt to take the *Tractatus* as metaphysical in a straightforward sense (as in Norman Malcolm's *Nothing is Hidden: Wittgenstein's Criticism of his Early Thought*, Oxford: Blackwell, 1986) yields plain nonsense or plain self-contradictions' (Diamond, *The Realistic Spirit*, *op. cit.*, n. 4, p. 19).

6 Ludwig Wittgenstein, *Tractatus Logico-Philosophicus*, trans. D.F. Pears and B.F. McGuinness (London: Routledge, 1922). Referred to throughout text as TLP.
7 James Conant, 'Putting Two and Two Together: Kierkegaard, Wittgenstein and the Point of View for Their Work as Authors', in T. Tessin and M. von der Ruhr (eds), *Philosophy and the Grammar of Religious Belief*, (Basingstoke: Macmillan, 1995), p. 250. Cf. also Ludwig Wittgenstein, *Philosophical Investigations*, ed. R. Rhees and G.E.M. Anscombe, trans. G.E.M. Anscombe, 2nd edn (Oxford: Blackwell, 1958), §§ 119, 464.
8 Conant, 'Must We Show What We Cannot Say?', *op. cit.*, n. 3, p. 246.
9 Conant, 'Putting Two and Two Together', *op. cit.*, n. 7, p. 249. *Prima facie* support is lent to Conant by Wittgenstein's remark that 'There is not a philosophical method, though there are indeed methods, like different therapies' (Wittgenstein, *Philosophical Investigations*, *op. cit.*, n. 7, § 133).
10 Conant, 'Putting Two and Two Together', *op. cit.*, n. 7, pp. 269–71.
11 See Diamond, *The Realistic Spirit*, *op. cit.*, n. 4, p. 19, p. 201.
12 Ibid., p. 182.
13 Richard R. Brockhaus, *Pulling up the Ladder: The Metaphysical Roots of Wittgenstein's Tractatus Logico-Philosophicus* (La Salle: Open Court, 1991, p. 1–2
14 Cf. Wittgenstein, *Philosophical Investigations*, *op. cit.*, n. 7, §§ 127, 89 and 90.
15 Marie McGinn, 'Between Metaphysics and Nonsense: Elucidation in Wittgenstein's *Tractatus*', *Philosophical Quarterly*, **49**, 1999, p. 501.
16 Cf. Bertrand Russell, 'On Denoting', reprinted in R.C. Marsh (ed.), *Logic and Knowledge*, London: Unwin, 1956, p. 54.
17 Cf. Donald Peterson, *Wittgenstein's Early Philosophy: Three Sides of the Mirror* (Toronto: University of Toronto Press, 1990), p. 48; Brian McGuinness, 'The Grundgedanke of the *Tractatus*', in *Understanding Wittgenstein*, Royal Institute of Philosophy Lectures, 1972/73, vol. 7 (London: Macmillan, 1976), pp. 49–50; Henry Finch, *Wittgenstein: The Later Philosophy: An Exposition of the 'Philosophical Investigations'* (New Jersey: Humanities Press, 1977), p. 5.
18 Pears tells us that Wittgenstein's 'picture theory' gave him '...a platform to criticise Russell's Platonic Theory of logic. The platform is Aristotelian in this sense: it brings down sentential forms from Russell's transcendent world and treats them as immanent in this world': David Pears, *The False Prison: A Study of the Development of Wittgenstein's Philosophy*, Vol. 1 (Oxford: Clarendon Press, 1987), p. 29.
19 Cf. Peterson, *Wittgenstein's Early Philosophy*, *op. cit.*, n. 17, p. 47.
20 Cf. ibid., pp. 36, 40; also Pears, *The False Prison*, *op. cit.*, n. 18, p. 29.
21 Cf. Peter Hylton, 'Functions, Operations and Sense in Wittgenstein's *Tractatus*', in W. Tait (ed.), *Early Analytic Philosophy* (Chicago: Open Court, 1997), p. 97.
22 Ibid., p. 96.
23 Bertrand Russell, 'The Philosophy of Logical Atomism', reprinted in R.C. Marsh (ed.), *Logic and Knowledge* (London: Unwin, 1956), p. 270.
24 Ibid.
25 Ibid.
26 McGinn, 'Between Metaphysics and Nonsense', *op. cit.*, n. 15, p. 497.
27 Ibid., p. 500.
28 Ibid., p. 148. One important difference between McGinn's reading and mine, is that she adopts the traditional reading of Wittgenstein's views on analysis and elementary propositions, whereas I do not. See McGinn, 'Between Metaphysics and Nonsense', *op. cit.*,

n. 15, p. 498. Cf. Lynette Reid, 'Wittgenstein's Ladder: The *Tractatus* and Nonsense', *Philosophical Investigations*, **21** (2), pp. 109–12.

29 Reid, 'Wittgenstein's Ladder', *op. cit.*, n. 28, p. 122.

30 Reid gives a useful account of Wittgenstein's views on issues concerning the end and nature of philosophical analysis that she contrasts with what he had written in the *Notebooks* (Ludwig Wittgenstein, *Notebooks, 1914–1916*, ed. G.H. von Wright and G.E.M. Anscombe, trans. G.E.M. Anscombe, Basil Blackwell: Oxford, 1969). See Reid, 'Wittgenstein's Ladder', *op. cit.*, n. 28, pp. 110–11. For example, whereas some examples of elementary propositions are attempted in the *Notebooks*, there are none in the *Tractatus*. She is right that one cannot simply move from what was written in the *Notebooks* to illuminate the *Tractatus*.

31 Cf. Wittgenstein, *Philosophical Investigations*, *op. cit.*, n. 7, § 134.

32 Wittgenstein, *Culture and Value*, *op. cit.*, n. 1, p. 7e.

33 Conant, 'Must We Show What We Cannot Say?', *op. cit.*, n. 3, p. 244. Conant stands opposed to those who maintain that '...for the *Tractatus* the propositions of ethics and religion – as well as either all or only the most important propositions of the *Tractatus* itself – are both nonsensical and deeply significant': ibid., p. 247. See also Conant, 'Putting Two and Two Together', *op. cit.*, n. 7, p. 252.

34 Conant remarks on the impossibility of there being different types of nonsense (cf. Conant, 'Must We Show What We Cannot Say?', *op. cit.*, n. 3, pp. 252, 253, 261). Diamond, for her part, says '...for Wittgenstein there is no kind of nonsense which is nonsense on account of what the terms composing it mean – there is as it were no positive nonsense' (Diamond, *The Realistic Spirit*, *op. cit.*, n. 4, pp. 106, 112). She goes on: 'I should claim that [this] view of nonsense...was consistently held to by Wittgenstein throughout his writings, from the period before the *Tractatus* was written and onwards' (ibid., p. 107).

35 Cora Diamond, 'Ethics, Imagination and the Method of Wittgenstein's *Tractatus*', in R. Heinrich and H. Vetter (eds), *Bilder der Philosophie, Wiener Reihe* **5** (Vienna: Oldenbourg, 1991), p. 65.

36 Koethe notes, 'On [Diamond's] account, the *Tractatus* does *not* attempt to articulate a metaphysical and semantic theory of the nature of language...Wittgenstein's aim is to subject the notions figuring in that ostensible articulation – states of affairs, objects, logical form, and so on – to "a destablization from the inside" in a effort to demonstrate their literal incoherence': John Koethe, *The Continuity of Wittgenstein's Thought* (Ithaca, NY: Cornell University Press, 1996), p. 37.

37 Diamond, 'Ethics, Imagination and the Method', *op. cit.*, n. 35, pp. 57, 55.

38 Wittgenstein, *Philosophical Investigations*, *op. cit.*, n. 7, p. viii.

39 Ibid., §§ 89, 92, 96, 98, 105, 107–8, 114–15.

40 Dale Jacquette, 'Wittgenstein and the Colour Incompatibility Problem', *History of Philosophy Quarterly*, **7** (3), p. 353.

41 This period is of great interest precisely because, as Finch notes, 'Judging from the posthumously published *Philosophical Remarks* and *Philosophical Grammar*, Wittgenstein moved with almost agonizing slowness to the new position that language makes sense only as it is altogether intertwined with different kinds of activities' (Finch, *Wittgenstein: The Later Philosophy*, *op. cit.*, n. 17, p. 4).

42 Richard McDonough, 'A Note on Frege's and Russell's Influence on Wittgenstein's *Tractatus*', *Russell*, **3** (14), 1994, p. 42. Cf. also McGinn, 'Between Metaphysics and Nonsense', *op. cit.*, n. 15, pp. 2, 4, 9–10.

43 Mayer convincingly argues against the view that the numbering was meant to be in the style of a musical score or an intuitive rhythm. Indeed, she employs evidence from the *Prototractatus* to reveal that the sub-propositions of the *Tractatus* are for the most part, just as Wittgenstein claimed, meant as elaborations or comments upon earlier remarks. Thus, she likens the work to 'an architectonic construction' and claims that its numbering system '...reflects primarily a method of construction' (Verena Mayer, 'The Numbering System in the *Tractatus*', *Ratio*, **6** (2), 1993, p. 112). This is at least consistent with the idea that these 'statements' were meant as elucidations.

44 Cf. Daniel Hutto, 'Was the Later Wittgenstein a Transcendental Idealist?', in P. Coates and D. Hutto (eds), *Current Issues in Idealism* (Bristol: Thoemmes Press, 1996), p. 147.

45 Wittgenstein, *Philosophical Investigations, op. cit.*, n. 7, § 19.

46 Ibid., § 23.

47 Cf. Koethe, *The Continuity of Wittgenstein's Thought, op. cit.*, n. 36, p. 2.

48 Bertrand Russell, *My Philosophical Development* (London: Routledge, 1959), p. 114.

49 Anscombe attempts to illustrate this point when she writes 'I once bought toffees with the names of the flavours "treacle", "Devon cream" and so on printed on the papers and was momentarily startled to find one labelled "fruit or nut". It cannot be "fruit or nut", I said. It's fruit or it's nut. Any attempt to say what the truth-functional constants like "or" mean must fail: we can only shew it' (G.E.M Anscombe, *An Introduction to Wittgenstein's Tractatus* (Philadelphia: University of Pennsylvania Press, 1959), p. 164).

50 Wittgenstein, *Philosophical Investigations, op. cit.*, n. 7, § 656.

51 Ibid., p. 226.

52 Ibid., § 206.

53 Meredith Williams, *Wittgenstein's Meaning and Mind* (London: Routledge, 1998), pp. 222, 180.

54 Ibid., p. 210.

55 Williams notes that by stressing the social character of meaning and the importance of training, Wittgenstein produces '...a dynamic rather than a static account of the rule-governed practice' (ibid, p. 168).

56 Wittgenstein, *Philosophical Investigations, op. cit.*, n. 7, § 654. Cf. also idem, *Last Writings on the Philosophy of Psychology*, Vol. I, ed. G.H. von Wright and H. Nyman, trans. C.G. Luckhardt and M.A.E. Aue (Oxford: Blackwell, 1982), § 873.

57 Wittgenstein, *Philosophical Investigations, op. cit.*, n. 7, §§ 373, 371.

58 Wittgenstein, *Culture and Value, op. cit.*, n. 1, p. 10e.

59 TLP 4.0031; Philosophical Investigations, op. cit., n. 7, § 122. Cf. also Koethe, The Continuity of Wittgenstein's Thought, op. cit., n. 36, p. 60.

60 Wittgenstein, *Philosophical Investigations, op. cit.*, n. 7, § 109.

61 Ibid., § 108.

62 Ibid., § 15.

63 Ibid., § 49.

64 In 1948 Wittgenstein writes, '...Nothing is more important for teaching us to understand the concepts we have than constructing fictitious ones' (Wittgenstein, *Culture and Value, op. cit.*, n. 1, p. 74e).

65 Wittgenstein reminds us that, 'The language games are rather set up as objects of comparison which are meant to throw light on the facts of our language by way not only of similarities, but also of dissimilarities' (Wittgenstein, *Philosophical Investigations, op. cit.*, n. 7, § 130; cf. also § 122).

66 Ludwig Wittgenstein, *Zettel*, 2nd edn, ed. G.E.M. Anscombe and G.H. von Wright, trans. G.E.M. Anscombe (Oxford: Basil Blackwell, 1967), § 380.

67 Ibid.
68 Ibid., § 379. For a fuller discussion of this case and the use of these imaginary language games see Hutto, 'Was the Later Wittgenstein a Transcendental Idealist?', *op. cit.*, n. 44, pp. 128–29.
69 Wittgenstein, *Philosophical Investigations, op. cit.*, n. 7, § 383. This way of understanding his later approach makes sense of his wholesale rejection of the view of psychological language as representational which sponsors a mistaken picture of the reified mind. For a fuller discussion of this see Hutto, 'Was the Later Wittgenstein a Transcendental Idealist?', *op. cit.*, n. 44, pp. 467–68, 474–77; also Wittgenstein, *Philosophical Investigations, op. cit.*, n. 7, §§ 194, 196.
70 Wittgenstein, *Philosophical Investigations, op. cit.*, n. 7, § 316.
71 Ibid., § 305.
72 Ibid., § 322.
73 Wittgenstein, *Last Writings, op. cit.*, n. 56, § 41.
74 Wittgenstein, *Philosophical Investigations, op. cit.*, n. 7, § 583.
75 Ibid., § 340; Wittgenstein, *Last Writings, op. cit.*, n. 56, § 856. Or, as he famously writes: 'To repeat: don't think, but look...' (*Philosophical Investigations*, § 66).
76 Wittgenstein, *Philosophical Investigations, op. cit.*, n. 7, § 68.
77 Ibid., § 69; cf. §§ 79–80.
78 Williams, *Wittgenstein's Meaning and Mind, op. cit.*, n. 53, pp. 76, 177.
79 He writes in his undated letter to von Ficker, '...the point of the book is ethical. I once wanted to give a few words in the foreword which now actually are not in it, which, however, I'll write to you now because they might be a key to you: I wanted to write that my work consists of two parts: of the one which is here, and of everything which I have not written. And precisely this second part is the important one.' Taken from C.G. Luckhardt (ed.), *Wittgenstein: Sources and Perspectives* (Bristol: Thoemmes Press, 1996), pp. 94–95.
80 He claims that Wittgenstein's preface to the *Tractatus* misleads on the issue of nonsense by suggesting a fourfold schema, when it should describe a sixfold one. Cf. Peterson, *Wittgenstein's Early Philosophy, op. cit.*, n. 17, p. 7, and especially p. 8. Another good account of the various senses of nonsense in the *Tractatus* can be found in Cyril Barrett, *Wittgenstein on Ethics and Religious Belief* (Oxford: Blackwell, 1991), esp. pp. 24–25.
81 Diamond, 'Ethics, Imagination and the Method', *op. cit.*, n. 35, p. 84; cf. also p. 80.
82 Ibid., p. 74.
83 If they could, ethical statements could not be absolute. Cf. Ludwig Wittgenstein, 'A Lecture on Ethics', reprinted in *Ludwig Wittgenstein: Philosophical Occasions 1912–1951*, ed. J. Klagge and A. Nordmann (Indianapolis: Hackett, 1993), p. 39.
84 Wittgenstein, *Philosophical Investigations, op. cit.*, n. 7, §§ 499–500. On the basis of these remarks I would endorse Reid's claim that in the later writings, '...what it is to exclude something as nonsense can vary': Reid, 'Wittgenstein's Ladder', *op. cit.*, n. 28, p. 146.
85 Reid, 'Wittgenstein's Ladder', *op. cit.*, n. 28, p. 146.
86 As Koethe writes, '...we should expect this attitude towards [the domains regarded as nonsensical] to persist beyond the *Tractatus*. If on the other hand, their construal as nonsensical in the *Tractatus* is an artefact or consequence of the picture theory, then we should expect such concepts and discourse to survive the abandonment of that theory and to occur in his later writings in ordinary, unconceptualised forms' (Koethe *The Continuity of Wittgenstein's Thought, op. cit.*, n. 36, p. 39). Despite our different characterizations of the so-called 'picture theory', I think, as does Koethe, that it is the second scenario which is realized.

- Probl. intr. of D. (128)
- TLPs reminders. (129) 134 (But for H its not nonsense that reminds.)
- H doesn't get C&D (138) Take up for crit.
- Points about continuity clearly in the target area of J's crit. (138) 139
- Lang-games rules etc. as transcendental (138-140
- Continuity lies in the transc. view of lang., only the 'theory' changes. (140)
- W as a transcendental philosopher. (143)
- Profound forms of nonsense. (145)

Chapter 7

Saying and Showing: An Example from Anscombe

Cora Diamond

I Anscombe's Example and Some Questions

This paper is dedicated to the memory of Elizabeth Anscombe, from whose work I have learned so much. The criticisms which I make of her treatment of saying and showing rest on ideas which she made clear in her discussions of Wittgenstein's later philosophy;[1] I believe that a problem with her treatment of the *Tractatus* is that she did not see how those same ideas were helpful, indeed essential, in considering his earlier thought.

In her *Introduction to Wittgenstein's Tractatus*, Anscombe has an interesting example which she uses to criticize what Wittgenstein says about saying and showing.[2] Speaking of the proposition '"Someone" is not the name of someone', she says that it is obviously true, and that it is an 'admittedly rather trivial' example of a proposition which lacks true–false poles. She believes that Wittgenstein would have said that the proposition is an attempt to say something that shows – something that cannot be said. But she thinks that the example suggests instead that what 'shows' in his sense *can* (at least in some cases) illuminatingly be said, contrary to what he holds in the *Tractatus* about propositions which lack true–false poles. Although she considers the example to be rather trivial, it is, I should suggest, a good example to think about for a number of reasons, including the fact that it does not directly involve any of the more difficult notions of the *Tractatus*, like pictorial or logical form. It is sometimes useful to move forward an inch at a time; and Anscombe's example, precisely because it is relatively trivial, provides an opportunity to try to move forward a short distance without taking on the topic of 'saying and showing' as a whole.

Since I shall be examining her discussion of the example, I shall need to quote the passage in question first. It comes immediately after her discussion of a somewhat different example, intended to bring out that there are cases in which what shows, in Wittgenstein's sense, cannot be *informatively* said. It is at this point that she remarks that what shows in that sense can nevertheless be *illuminatingly* said, and she goes on to give the example of '"Someone" is not the name of someone'. She continues:

> This is obviously true. But it does not have the bipolarity of Wittgenstein's 'significant propositions'. For what is it that it denies to be the case? Evidently, that 'someone' is the name of someone. But what would it be for 'someone' to be the name of someone? Someone might christen his child 'Someone'. But when we say that '"Someone" is not the name of someone,' we are not intending to deny that anyone in the world has the odd name 'Someone'.
>
> What then are we intending to deny? Only a piece of confusion. But this *sort* of denial may well need emphasizing. Students, for example, may believe what Professor Flew tells us...: namely that 'somebody' refers to a person, that it is part of the 'logic' of 'somebody', unlike 'nobody', to refer to somebody. If this were so, then on being told that everybody hates somebody, we could ask to be introduced to this universally hated person. When we say '"Somebody" does not refer to somebody', what we are intending to deny is what Professor Flew meant. But he did not really mean anything (even if he felt as if he did).
>
> Here a statement which appears quite correct is not a statement with true–false poles. Its contradictory, when examined, peters out into nothingness. We may infer from this that Wittgenstein's account of propositions is inadequate, correct only within a restricted area. For it hardly seems reasonable to prohibit the formula: '"Somebody" does not refer to somebody' or '"Someone" is not the name of someone'; nor of course, is this logical truth in any sharp sense of 'logical truth'. It is, rather, an insight; the opposite of it is only confusion and muddle (not contradiction).
>
> The example of '"Someone" is not the name of someone' is particularly clear, because the true proposition is negative. According to Wittgenstein, however, since what our proposition denies does not turn out to be anything, it itself is *not* a truth; for there isn't anything which it says is not the case, as opposed to the equally possible situation of its being the case. Therefore Wittgenstein would either have looked for a more acceptable formulation (which I think is impossible) or have said it was something which *shewed* – stared you in the face, at any rate once you had taken a good look – but could not be *said*. This partly accounts for the comical frequency with which, in expounding the *Tractatus*, one is tempted to say things and then say that they cannot be said.

The case is less clear than Anscombe suggests. We can, to start with, note the ambiguity on which Anscombe herself comments in the first paragraph, after she asks what the utterer of '"Someone" is not the name of someone' intends to deny. Not, she says, that there is a person who has the odd name 'Someone'.

Now if the utterer's remark *could* have been used as a denial that there is a person with the odd name 'Someone', it should be possible to recast the remark itself to remove the ambiguity. But, before we consider that sort of ambiguity, we should note that the original remark is not ambiguous merely in the way Anscombe herself enables us to recognize. There is a further sort of ambiguity about which she does not comment. For the remark might be intended, as Anscombe indeed intends it, to have a general application, but its intended application might instead be much more limited. That is, someone might say '"Someone" is not the name of someone' in response to confusion about the role of 'Someone' in some particular utterance. A child, perhaps, has heard at school that someone has measles, and thinks that what has been said concerns a person called 'Someone': that person has measles. The remark '"Someone" is not the name of someone', spoken to the child, may then be about the word 'Someone' as it occurs in 'Someone has measles'. This is different in significant respects from the case in which the speaker of '"Someone" is not the name of someone' is responding to confusion of the sort that might be induced by Professor Flew. The response to Flew is not directly about any specific occurrence of the word 'Someone'. The speaker means to speak about all its uses. – But is that right? Surely not, for, as Anscombe notes, the remark isn't meant to cover cases in which we are concerned about whether anyone has the odd name 'Someone'. So what sort of generality does the remark about 'Someone' not being a name have, when it is meant to have general applicability? Is the point that 'Someone', whenever it hasn't got the logical role of a name, isn't a name? That won't do; for, understood in that way, the remark no longer appears to be something *illuminating* – something that might be called an *insight*. And Anscombe insists that the point she takes herself to be expressing is not a logical truth in any sharp sense of the term. Whatever exactly she means to include in 'logical truth in a sharp sense', it would presumably encompass the kind of truth that might belong to '"Someone", when it isn't used as a name, isn't used as a name'.

I hope to have shown that Anscombe's example needs some clarification. But before attempting to provide it, I should comment that the issues here are directly connected with questions about how to read the *Tractatus*. Anscombe evidently believes that at least some of the propositions that Wittgenstein does not recognize to be significant propositions are significant; they cannot, though, be recognized as significant if one accepts the *Tractatus* account. She would not take these propositions to be logical truths in a sharp sense, any more than she takes '"Someone" is not the name of someone' to be logical truth in a sharp sense. What I shall be trying to show is that, when we are clearer about the kind of clarification that would be helpful in the case of the confusion that she has described, we shall see that the activity of clarification can, in some cases, be thought of as adding tautologies to the propositions that need clarification, and contrasting them with propositions to which somewhat different tautologies have been added. While this process can make clearer how we are using a sentence, it does not have the character which Anscombe takes clarification to have. What I intend to demonstrate in this paper is that we need to clarify philosophical clarification if we are to think clearly about the question how the *Tractatus* is to be read, the question whether it should be read as containing propositions which are genuinely significant but which could not have been recognized as significant by its author because of the doctrines which he accepted when he wrote it. One case cannot show us how to read the book; it can show us some of the things we need to think about.

I should add that this essay is directed also against an idea of John Koethe's, that it is an objection to a reading of the *Tractatus* as not genuinely intended to convey doctrines through its propositions, that that reading leaves us dealing with the differences between sense and nonsense piecemeal.[3] My point against that is:

right! That is, the only thing wrong with the idea is that Koethe takes it to be an objection. It is, indeed, by working through what is involved in a piecemeal approach to philosophical clarification that we can see how Wittgenstein thought of the activity of philosophy. That is, I want to suggest that it is no accident that Anscombe goes wrong in her discussion of '"Someone" is not the name of someone'. She moves directly from the proposition's supposedly saying something true and its lacking true–false poles to its failing to count as a significant proposition on the *Tractatus* view. (Indeed, she describes the proposition as counting, from the point of view of the *Tractatus*, as a 'prohibited formula', a description at odds, at least apparently, with Wittgenstein's claim that no combination of signs is logically impermissible, which seems to mean that there is no such thing as a 'prohibited formula.') Her own view is that the proposition does connect with what the confused person, in this case Professor Flew, took himself to have in mind. So she does not look at the details of how it might be supposed to connect. My argument is that only if we do go into the details can we see the status of the proposition. We have to look at its use from close to.

In *Philosophical Investigations* Wittgenstein speaks of our reluctance to look at the use of words 'from close to' – our reluctance to consider the details. This is certainly meant as a criticism of his own former approach. But our recognition that it is meant as such a criticism may make it hard for us to see the complexity of the case, and may make it difficult for us to see what exactly was involved in his earlier method. It may make us think that the earlier method was meant to enable us to show that some proposition or other was nonsensical with a mere wave of the hand. So Anscombe thinks that her own proposition simply turns out to be nonsensical by the application of a general *Tractatus* principle, and the issue of what its use *is* – of what she really wants to do with it – is not examined. She knows in her bones that she means something – something illuminating, though lacking true–false poles – so, although she also knows that the sense one may have of meaning something when one speaks is capable of misleading one (she says that Flew is misled in just that way), she does not turn a careful analytical eye on to the details of what she wants to say. Here, I am trying to point to a kind of parallel between Wittgenstein's earlier and his later philosophy, a parallel in the role of forms of reluctance to look at details of use. Wittgenstein's understanding of clarification, early and late, goes with conceptions (somewhat different, early and late) of what, in philosophy, we may be reluctant to do.

II An Easier Case

It will be useful to consider, first, the kind of case in which the remark '"Someone" is not the name of someone' is intended to apply to a particular utterance. Although this is not the kind of case Anscombe herself has in mind, our discussion of it will be made easier if we approach Anscombe's kind of case after considering this other case.

Details "from close to"

We are imagining, then, that what has been said is 'Someone has measles', and a child is confused about what that means. The confusion may have been shown in her attempt to treat 'Someone' as a name; perhaps she shows by something she says that she takes it that the proposition's falsity would be inferable from the falsity of a proposition about some one individual. If *that* child has not got measles, then 'Someone has measles' would have to be false, or so she thinks.

Now we can make clear her mistake by making it explicit that the role in inferences of 'Someone has measles' is different. If we are talking about some specific group of possible measles-havers and measles-not-havers, we can make it explicit that only from the premise that each and every one of them has not got measles could we infer the falsity of 'Someone has measles'.

Anscombe herself imagines explaining to the confused person that 'Someone' does not figure in multiple quantification as a name of someone. In the kind of case I am imagining, in which the confusion concerns the role of 'Someone' in a particular utterance, the point about multiple quantification is, as it stands, not directly helpful (even if it could be put in language plain enough for the confused person to follow). For, if someone thinks that 'Someone' in 'Someone has measles' is the name of some particular person, she won't be helped by being shown that 'Someone' in 'Everyone hates someone' is *not* a name, unless she is given some explanation how that is relevant to its use in 'Someone has measles'. Perhaps the confused person would be helped by the discussion of 'Everyone hates someone' if it helped her to drop an assumption that 'Someone' in *every* context had to be a name. Anscombe would, I think, hold that 'Someone' has a single characterizable use such that both 'Someone has measles' and 'Everyone hates someone' contain 'Someone' used in that way. This, too, is fine; but what it means is that the bearing of the use of 'Someone' in 'Everyone hates someone' on the use of 'Someone' in 'Someone has measles' cannot be explained without bringing in what it is for these sentences to contain 'Someone' used in the same way.

Consider a particular kind of case, one of the kinds of case that lie behind Flew's remark.[4] Imagine that someone, Jane, has said 'James loves somebody'; later, on introducing Alice, Jane says 'When I said that James loves somebody, I was referring to Alice'. Flew (who was, after all, concerned to contrast 'Somebody' and 'Nobody') may have wished to emphasize the possibility of such remarks, which are impossible in connection with 'Nobody' in its usual use. If this kind of case is behind the confusion which Anscombe wants to address, the problem mentioned in the last paragraph plainly arises when we try to help the person suffering from the confusion by noting that there is, in the case of 'Everybody hates somebody', no appropriate demand that one be introduced to the person referred to – in the sense of 'referred to' in which Jane had said that she had referred to Alice. The difficulty can be explained in terms of the ways in which we may think about the differences from each other, and similarities to each other, of the three propositions: 'James loves Alice', 'James loves somebody' and 'Everybody hates somebody'. A would-be helper of someone in the confusion induced by Flew can show us, by explaining the behaviour of the third proposition, that we cannot treat the middle proposition as logically like both the first and third propositions. But, unless we are helped to see that 'Somebody' in the third proposition exemplifies the *same* use as 'Somebody' in the second, we may remain overimpressed by just those similarities between the first and second propositions which lead into the original confusion.

One of the problems with Anscombe's clarification is that she takes it to be useful to point out that 'Somebody' doesn't refer to somebody, taking the word 'refer' not itself to be in need of clarification. But, if we take seriously the kind of example I have just described, we can see that 'refer' has a use in connection with 'James loves somebody'. I don't, in using that proposition, refer to Alice by 'Somebody' in the sense in which I refer to someone when I use a name for that person, but there is a perfectly respectable way of using 'refer' in connection with the remarks made by Jane. (I shall return to this point in Section III.) We can also see that Anscombe's point, that Flew did not really mean anything, although he may have felt that he did, is too strong. For he sees plainly enough that Lewis Carroll's jokes about 'Nobody' in *Through the Looking Glass* would not be jokes about 'Somebody'. (See the quotation from Flew in Section III below.) Flew uses 'refer' to explain the difference, without seeing any need to explain any contrast between a use of 'refer' which can be connected with some occurrences of 'Somebody' working as a variable and other ways of using 'refer'. Anscombe's claim that Flew has nothing in mind depends on her sharing with Flew the idea that the word 'refer' can easily bear the kind of weight we may want to put on it in our clarifications. That it cannot do so is shown by the way in which Flew fails to make clear what he means and instead says something which easily leads into confusion, and by the way in which Anscombe fails to see that he *does* mean something.

On Wittgenstein's view, we cannot make fully clear the use of a word in one proposition if we are not clear what *further* uses of the word would be uses of it in the same way. So the issues raised by Anscombe's imagined use of 'Everybody hates somebody' cannot be set aside. However, we can temporarily set them aside and focus simply on what might be involved in removing a confusion about the occurrence of 'Someone' in a particular utterance.

In fact, it is useful to consider the way in which the clarification of the particular utterance can involve a point related to that made by Anscombe when she brings in multiple quantification, without actually bringing in multiple quantification. For we should consider the way in which the child's confusion about 'Someone has measles' might show itself in inferences from 'Someone has measles and someone has mumps' to 'Someone has measles and mumps'. (I simplify my discussion of the example by leaving out qualifications for odd cases such as those in which someone takes 'measles' and 'mumps' to be two names for the same disease.) To infer in that way is to take 'Someone' to be a name; but it is possible to take 'Someone' to be a name, and nevertheless not to regard the inference as available. One might, for example, take 'Someone' in 'Someone has measles' to be a name of one person, and 'Someone' in 'Someone has mumps' to be the name of someone else. So the point about the unavailability of the inference to 'Someone has measles and mumps' might help the child but it might not; for it leaves it open that

'Someone' is a name – a name of more than one person. If it does not help, one might pursue the point, assuming a child with a rather unnaturally good uptake for logic, with 'Each of measles and mumps is had by someone' and its not allowing (if it is understood as one wants it to be understood) the inference to 'Someone has measles and mumps'. But what we need to see here is that the process of clarification involves making clear as much of the inferential behaviour of the utterance with which we are concerned as is necessary.

I have been describing how an explanation of the inferential connections of 'Someone has measles' may enable a child to grasp the role of 'Someone' in it. Here we can note a couple of *Tractatus* points. First, if you add a tautology as a conjunct to a senseful proposition, the result will be identical with *the original senseful proposition*: you have what counts, from the point of view of the *Tractatus*, as the same proposition (see TLP 4.465).[5] Secondly, in the *Tractatus*, inference is fundamentally tied to tautologies. To any valid inference there corresponds a tautology.[6] Logical propositions are tautologies and they are forms of proof: 'they shew that one or more propositions *follow* from one (or more)'.[7] If you can infer some conclusion from some set of premises, you can construct a proposition which has truth-functional 'implies' as its main connective, using the conjunction of the premises of the inference as the antecedent, and the conclusion of the inference as the consequent; you will have a proposition that can be seen to be a tautology. That it is a tautology can be mechanically determined, although making clear that it is a tautology involves something which cannot be carried out mechanically, namely the working out of the analysis of the component propositions far enough for the tautological character of the combination to become evident. Further, the mechanical demonstration that a combination is tautological depends on taking signs that recur as recurring with the same meaning. In TLP 4.1211 Wittgenstein says that two propositions '*fa*' and '*ga*' show that they are both about the same object; but they show this only in that '*a*' is being used in the same way in both. But, for *that* to be something that shows, we must be using '*fa*' and '*ga*' so that we can infer that there is something that is both *f* and *g* from the two propositions together. For a combination to be a tautology and for us to be using the component symbols in inferences in certain ways are the same thing. Nevertheless we can clarify a proposition, help make clear to someone what proposition it is, what its use is, by writing it with a particular helpful tautology added to it.[8]

I have just claimed that, on the *Tractatus* view, there is a tautology that corresponds to a valid inference, and hence that the use of a proposition in inferences can be clarified by adding a helpful tautology, corresponding to the valid inference, to the proposition. That one proposition does not follow from some set of propositions is also reflected in a tautology. One can write out the truth-functional implication corresponding to the non-allowable inference, and all the non-tautological combinations of all the propositions which are components of the premises of the non-allowable inference and of the proposition giving what cannot be inferred from them, carrying through the analysis of the various propositions as far as is necessary. There will be a tautological equivalence between the truth-functional implication corresponding to the inference and one of the non-tautological combinations. That tautological equivalence of the implication with the non-tautology reflects the invalidity of the inference. Wittgenstein made a closely related point later on, when he said that we do logic by proving that certain propositions are tautologies, but we could do it as well by proving that certain propositions are not tautologies.[9] (That a combination is non-tautological depends on the use of the symbols, just as does the fact that some particular combination is a tautology.)

The two *Tractatus* points – namely (1) that if you add a tautology to a proposition, the result is the original proposition, and (2) that both the availability of some inference and the unavailability of some inference can be made clear by writing tautologies – mean that any clarification of a proposition which works by clarifying its inferential behaviour can be thought of as adding a tautology to the original proposition; and this is to say that any such clarification could be described as adding nothing to the original proposition. It simply rewrites it, giving us the same proposition, expressing the same thought, with fundamentally the same mode of use of signs; it rewrites it, though, in a way that may remove unclarity.[10]

Let us apply this point about the character of clarification to the case of the confusion about 'Someone has measles'. Write down the sentence 'Someone has measles' twice. Add to the first a specification of one kind of inferential behaviour; add to the second a specification of a different pattern of inferential behaviour. By this I mean the two patterns of inferential behaviour discussed above – that is, the use of 'Someone has measles' so that its negation can be inferred from a single proposition about one person not having measles, and the use of 'Someone has measles' so that its negation cannot be inferred from any premise short of one denying of every member of the group in question that she has measles. The second use also disallows the inference from the two premises 'Someone has measles' and 'Someone has mumps' to 'Someone has measles and mumps' and so on. To specify the two patterns of inferential behaviour is to do something that could also be done by adding one group of tautologies to the original sentence, to clarify one use, and another group of tautologies to the sentence, to clarify a distinct use. It is worth emphasizing here that no tautologous addition does the work of clarification except by being picked up, and the 'picking up' in question involves the very matters being clarified – that is, the uses of 'someone'. From the point of view of logic, any tautology is equivalent to any other. So how can one tautological addition help to clarify one proposition, and another clarify a different proposition? The answer is: by being taken as a helpful guide to use; in other words, a proposition rewritten with the tautology in question as part of it may be seen in a certain way. It may be seen with its use more clearly in focus.[11]

Wittgenstein held not only that there is a sense in which we cannot go wrong in logic, but also that we cannot specify the sense of propositions in some wrong way. This means that neither of our specifications of inferential behaviour for 'Someone has measles' contains any logical error, but we might well find that, if we in fact used the sentence in these two different ways, people would run into practical problems; having both the uses would be inconvenient. We might be well advised not to opt for a use which allows the inference from the negation of a proposition saying of some particular person that he has measles to the negation of 'Someone has measles' – that is, not to use 'Someone' as a name. ('Someone' and 'Somebody' may sometimes be used as nonce-names, as 'Nobody' occasionally is – for example, by Chesterton's Father Brown, who speaks of 'Nobody's glass' in drawing attention to the presence of a glass of whisky drunk by someone not yet considered by the investigators of a murder. 'Nobody's glass' does not there mean a glass that isn't anybody's; rather, the word 'Nobody' is a nonce-name for the man otherwise identified as 'The Quick One'. 'Somebody' and 'nobody' are also used as predicates meaning someone of some consequence and someone of no consequence.)

Suppose, then, that you have specified the two patterns of inferential behaviour, and that you explain them to the person who was confused about 'Someone has measles'. If she sees that there are these two different uses, her confusion may disappear. You may, if you like, say: 'When there is *this* type of inferential behaviour, I call "Someone", used this way, a name, and if there is *that* type of inferential behaviour, I say that "Someone", used that way, is not a name.' But, then, if that is your explanation of 'is a name' and 'is not a name', all you would be doing if you later described a use of 'Someone' as not a use as a name is adding a specification of the inferential behaviour of the propositions you are speaking about to those propositions; and on the *Tractatus* view that is doing the same as adding tautologies to the propositions in question. What is more important is that the use of the label 'name' is not what is significant in the clarification, since the removal of confusion is entirely a matter of seeing the differences in the inferential patterns, and not a matter of applying any label to one use rather than the other.

So, in connection with the kind of case we are considering – namely a confusion about the role of 'Someone' in a particular utterance – what can illuminatingly be *done* is a clarification of the inferential behaviour of the proposition. This activity can be regarded, on the *Tractatus* view, as in effect the addition of tautologies to the proposition being clarified, perhaps with contrasts to other propositions with tautologies added to them. Clarity about clarification in this kind of case thus raises the question whether there is room in such cases for the sort of point in which Anscombe is interested – a point expressed in a proposition which is illuminating but which lacks true–false poles and is not a mere tautology.

Although we are not yet dealing with the kind of case Anscombe has centrally in mind, we should turn back here to the final paragraph of the quoted passage from Anscombe. She says there that she does not believe that there is a more acceptable formulation (from the point of view of the *Tractatus*) of the point made by saying that '"Someone" is not the name of someone', and she contrasts that with what Wittgenstein seems to be inviting us to do – namely, look hard at something which supposedly stares us in the face. I have not argued that there is some more 'acceptable' formulation of Anscombe's kind of point, available to us in the case of the confusion about 'Someone has measles', but rather that the sort of proposition which she takes to be illuminating is not what does the work of clarification. To clarify in the relevant way, in this kind of case, is to make plain the use of our expressions; this is not done by saying of such-and-such a sign that it is or is not being used in such-and-such context as a name, unless we are already able to connect 'is being used as name' with a pattern of use. It is the capacity to make the connection with the pattern of use that is essential. But the signs in question, in this case the sentence 'Someone has measles', can be used in various ways; we cannot say that 'Someone' in the mere sentence isn't a name. It might in some use be a name. What we want to speak of is 'Someone has measles' used in a particular way. But if we succeed in making clear what use of the sentence we are talking about, we have already made clear how the word 'Someone' is being used in it; to say that it is not a name would not add anything to the clarification we had achieved of what use of the sentence we had in view. (My own earlier reference to someone who thinks that 'Someone' is being used as a name of two different individuals has to be taken in the way I have just specified. If my description can be followed, it is because the person who can follow it knows how to discriminate between two different patterns of inferential behaviour, which might be labelled as 'use as name of one individual' and 'use as name of two individuals'.)

Anscombe speaks of Wittgenstein's supposed view that all we need to do is look hard at what stares us in the face in order to see that 'Someone' is not a name. Well, suppose that Wittgenstein would indeed have said that the proposition 'Someone has measles' shows that 'Someone' in it is not a name. This is hardly to say that we have only to look hard at the sentence to see that 'Someone' is not a name. What would 'looking hard' be? We need to look at signs with their use, and, while it might in some sense be true to say that this 'stares us in the face', it isn't obvious in what sense it is supposed to do so. But Wittgenstein did think that there was an activity through which a proposition can be turned into a new version of itself; the activity helps us grasp what we might then realize had shown all along (had shown in that the proposition had such-and-such determinate use). The activity of clarification turns propositions into versions of themselves that enable us to see clearly what the propositions had in a sense shown all along. I believe that some such account of clarificatory rewriting can be seen to be involved in Wittgenstein's conception of analysis – analysis being a form of clarificatory rewriting tied to the principle that, if p follows from q and q from p, they are the same proposition. But I cannot here attempt to give an account of Wittgenstein's view of analysis.

Philosophy, the *Tractatus* says, aims at the logical clarification of thoughts. It is an activity which results in propositions getting clear, not in philosophical propositions. By giving us new versions of our propositions, it takes thoughts that were not in focus, and makes it possible for us to avoid the confusions that came from that lack of focus. I hope that this section has demonstrated how it is (in some cases) possible to get that shift from lack of focus to focus by doing something that we can, from within the point of view of the *Tractatus*, regard as adding tautologies to our propositions.

I said that the activity of clarification turns propositions into versions of themselves that enable us to see clearly what the proposition had shown all along. This description of the activity is meant to connect with a problem about the translation of TLP 4.112. In a letter to Ogden about this passage Wittgenstein wrote that he thought 'it cannot be the RESULT of philosophy "to make propositions clear": this can only be its task. The *result* must be that the propositions *now have become clear* that they ARE clear'.[12] This remark is almost bound to be misleading if we do not bear in mind that very different ways of writing a proposition will give us what still counts as the same proposition; and what proposition the proposition is, what symbol it is, may be far easier to take in in some modes of writing than in others. A Russellian analysis, for example, rewrites propositions in a way that might be said to make clear that they 'were' clear. That is, the sense in which they 'were' clear is that they were all along the same proposition as the proposition now written in a way which eliminates certain possibilities of confusion. To say, as I did, that philosophy results in propositions getting clear may seem to contradict Wittgenstein's point in his letter to Ogden. That is, it looks as if I am saying, against Wittgenstein, that the propositions in question were not already clear. And the response (if someone made that objection to me) would be to say that, on his view, in one sense they were already clear and in another sense they weren't. The sense in which they weren't is that we have work to do before their clarity (in the sense in which they have it all along) is available to us. This leaves open the question how far we need to push the work of clarification.

III Back to Anscombe

In Section II, I mentioned that there is a respectable way of using the word 'refer' in which some occurrences of 'somebody' or 'someone' can be said to refer to someone, as when I ask you 'When you said that someone had made you very angry today, were you referring to the Dean?'. Let us look at Flew's example and his remarks about it – the remarks that led Anscombe to say that Flew meant nothing, even if he thought he meant something. Flew quotes Lewis Carroll:

> 'I see nobody on the road,' said Alice.
> 'I only wish *I* had such eyes,' the King remarked... 'To be able to see Nobody! And at that distance too! Why, it's as much as *I* can do to see real people by this light!'

Flew says that the King's mistake is to treat 'Nobody' as if it had the logic of 'Somebody', as if 'Nobody' referred to somebody, albeit a rather insubstantial somebody.[13] Flew's confusion, if he is indeed confused, does not lie in the first part of his claim, about the King having treated 'Nobody' as if it had the logic of 'Somebody'. What he means by its having the logic of 'Somebody' is clear in the quotation from Carroll: you treat 'Nobody' as having the logic of 'Somebody' if you take it to be possible to infer from Alice's 'I see nobody on the road' that you also would be able to see whoever it is Alice sees, if you had Alice's capacity to see. If Alice sees someone, you ought to be able to see that person if you had eyes as good as hers. (And, since you can, although perhaps just barely, see all real people in the available light, the person whom Alice sees and whom you can't see can't be a real person.) To think in that way is to treat 'Nobody' as having the logic of 'Somebody'. When Anscombe argues that Flew was confused, she says that, if he were correct that 'Somebody' refers to somebody, we could, if told that everybody hates somebody, ask to be introduced to the universally hated person. But what stands behind the use of 'refer' in Flew's remark about the Carroll example is that, in the case of being told 'Alice sees someone', we could ask to be introduced to the person whom she sees, and we might indeed expect to be able, in appropriate circumstances, to be able to see that person ourselves. So, if all that Professor Flew meant was that that sort of move was possible in the case of saying 'I see somebody' but not possible in the case of saying 'I see nobody', there would be nothing at all confused in his remark. (He does not at any point say that 'Somebody' is a name, and it is questionable whether his remarks suggest that it is.) Anscombe, who quotes only a part of Flew's discussion, takes him to intend also to be speaking about a range of other cases in which 'Somebody' is used including uses of 'Everybody hates somebody' (but not including cases in which 'Somebody' is actually used as somebody's name), and to intend the word 'refer' in a particular sense, in which to say of a word that it refers is to say that it functions as a name.

If my arguments in Section II are correct, it would follow that the work of clarification of use is not in general done merely by description of a use of a word as a referring use, or as the use of the word as a name, or as a variable: behind such a description there has to be a grasp of different patterns of inferential behaviour. This is particularly clear in the case of 'refer'. It can indeed be misleading to say that 'Somebody' is a referring expression, but there is nothing particularly misleading about saying 'When I said that I was angry at somebody, I was referring to the Dean'.

There is, here, a methodological point about demonstrating that what someone has said is nonsense, as Anscombe wants to do in Flew's case. For, if an utterance appears to be nonsensical at first, there may very well be some not-nonsensical way of taking it. In the case of Flew's remark, it is plain that there is a way of taking his quoted example, together with his remarks about confusing the logic of 'Nobody' with that of 'Somebody' and about thinking that 'Nobody' referred to somebody as 'Somebody' would, so that what he says is not nonsensical but uses 'referred' in a way different from what Anscombe takes as the use of that word. If she takes what he says to be nonsensical, it is (I think) because she is convinced that she sees what he is trying to mean, and she takes *that* to be mere confusion. Taking his use of 'referred' as she thinks the word must be taken, she reads Flew as saying what could also be put as '"Somebody" is a name of somebody', and, while that could (on the view she takes in the passage quoted) mean something true if there were indeed someone called 'Somebody', what Flew intends is, she thinks, something else – something that is mere confusion. I believe that the idea of a nonsensical sense plays a role in her reading (or what I take to be her misreading) of Flew. If one believes that one had hold of what a person is trying to say and that *that* is nonsense, is no meaning at all, there is almost a guarantee that one will not attend to possibilities of sense of the sort that can be obscured if there is some unsuspected difference in the way a word is being used. If one wanted to show that what someone had said was indeed nonsense, one would need to go about it differently: one would need to think through how his words could perhaps be taken this way or that way, etc., and one would need, then, to bring out why one took it that none of the possibilities would answer to his intentions. Obviously here there can be no conclusive demonstration, for the person might well be able to show that the words had been intended in some other way.[14]

I have argued that there are problems with Anscombe's attempt to show that Flew was confused. I want now to turn to another matter: her claim that we can illuminatingly say '"Someone" is not the name of someone', and that this indicates that we can indeed say the kind of thing that Wittgenstein took to be showable but not sayable. We have seen, though, that it is not clear what (if anything) is said by '"Someone" is not a name of someone' unless, at the least, it is clear what use or uses of 'Someone' the person making the remark means to be speaking about. Anscombe herself recognizes this, but does not, I think, follow through on the consequences. We can lay out this or that use of 'Somebody'; this would involve the same kind of activity that Anscombe herself engages in in connection with the use she has in mind of 'Everybody hates somebody'. Her intention is to criticize Flew's description of 'Somebody' as referring to someone; her talk of whether it is or isn't a name is meant to make clearer what she is objecting to. She gives us part of a specification of what it is to use a word so that it could be said to refer; but suppose she were to lay this out more clearly. And suppose she were also to lay out the use of 'Somebody' that she has in mind. Could it illuminatingly be said that that use was not the sort of use that she took to be properly describable as a referring use? If so, in what way is it illuminating? For to say it is not a referring use will be to distinguish it from the use she has laid out as what she wished to call a referring use. But the difference between the use she had laid out as 'referring use' and the use she had laid out as the relevant use of 'Someone' (the use about which she wants to say something) will be as evident as it can be quite independently of any label for the uses. It may be illuminating to point to one thing, and to point to another, and to say, 'Look!' – or, in this case, to give a specification of one sort of use, and of another, and to say 'Look!' – but Anscombe's conception of what she is engaged in is meant to be contrasted with any such directive; it is meant to be a saying of something true that it is not a logical truth in any sharp sense of logical truth. But that a mode of use that allows such-and-such inferences (and so on) is not a mode of use that disallows those inferences (and so on), if it is any sort of content, would not be the kind of content she takes her remark to have. Here I want to emphasize that I am not disputing that a remark of the sort she makes may, in some contexts, be illuminating; the question is whether, when it is illuminating, it does anything different from specifying, to the extent that it is helpful, some use or uses and saying 'Look!'.

If I am right, then, Anscombe takes herself to be conveying something with a genuine content, although there is no content of the sort she takes herself to be conveying. Attempts at spelling out what is meant by her remark would involve laying out uses in such a way that their differences from other uses were plain. If, as I think, her own account is beset by problems, these arise in part through a blurring of Wittgenstein's distinction between sign and symbol, and in part through the use of formal concepts as if they were ordinary concepts. When Anscombe says that '"Someone" is not the name of someone' she immediately makes the move to eliminate an irrelevant use of the *sign* 'Someone', its use as an honest-to-goodness name. But her remark nevertheless retains its appearance of content largely because she does not specify the use of 'Someone' about which she *does* want to speak. She does not enable us to have the symbol in question, the symbol about which she wants to speak, clearly in view. But if the symbol were made clear, were put clearly before us, the appearance of there being something that she wants to say about it – something to be said which is not logically true in any sharp sense – would be harder to achieve. It may be helpful in understanding the issues here to note that 'Somebody' and 'Nobody' are frequently used in ordinary language as predicates. When Anscombe says '"Somebody" is not the name of somebody', she does not mean 'Somebody' as used in 'He really thinks he's somebody', just as Flew does not mean to talk about 'Nobody' as used in 'His people are nobodies'. Each of them intends to speak about a different symbol; and, if it is necessary to clarify their remarks, clarification will involve making plain what symbol they mean. One doesn't have to think of cases in which 'Somebody' or 'Nobody' might actually be used as a name to make clear that the purported subject matter of their remarks is not a sign. Again, 'is a name of someone' is, on the *Tractatus* view, a characterization of a mode of use of signs; the expression for it, in a logically perspicuous symbolism, is a variable, the values of which are propositions containing signs used in the relevant way. There need be nothing the matter with talk of something's being or not being a name, or of some sign's being or not being a name of two distinct items, or whatever; but the appearance that such remarks may have, of saying something that goes beyond what Wittgenstein allows to be sayable, is created in part by not looking behind the label 'name' or 'name of two things' to the pattern of use. At TLP 4.1211, Wittgenstein says that the proposition '*fa*' shows that the object *a* occurs in its sense. This obviously makes it look as if what is shown is something that can be put into a 'that' clause. And since what follows the word 'that' does not count as sayable, it looks as if we have the comical case Anscombe refers to, of remarks that say what it is that cannot be said. The crucial thing we need to attend to here is not what comes after the 'that' clause but what comes before it. For what is it that supposedly does the showing? Signs used in a certain way. But what way? To lay out the use in question, including the inferential behaviour of the proposition, would let us see the proposition itself more adequately. In this case, it would include being able to see clearly the use of the sign '*a*'. If we see the proposition itself clearly, there is then nothing at all *further* to be had, nothing further that might be said or thought or conveyed in some wordless way or even left unsaid.[15] That a proposition shows its sense means only that to see its use is to see what it says is so. The seeing here isn't a grasping of an inexpressible something that can 'only be shown', but a logical capacity, a capacity for intelligent use, for use in accordance with the rules of logical syntax.[16]

IV Conclusions

I have not argued that Anscombe's remarks about 'Somebody' and 'Someone' are nonsense. I would argue that, if her claim is that there is something she means that is not tautologous, not logical truth in a sharp sense, it is not at all clear what this might be. The difficulty is that of finding something that is what she wants to say, given that any clarification that turns it into a tautology or anything like a tautology will not be what she wants. She does not want to speak merely about the signs 'Someone' and 'Somebody', divorced from their use as ordinary-language variables. For, if we talk about the mere signs, what would be meant by denying that they were names of anybody would be merely that they had not been given a particular kind of use: no one has been given the name 'Someone' or the name 'Somebody'. And that is not her point. (I would think it likely, in fact, that the name 'Somebody' had been borne by some cat or dog.) On the other hand she cannot be said to have wanted to talk about this or that particular symbol, or at any rate not clearly to have wanted to do so. For if she had wanted to say something about a symbol, she would have had to lay out what use of the sign she had in mind – that is, what symbol she was talking about. And then the question would be what she wanted to say about the symbol that had not already been made clear in giving what symbol it was. I have argued that the impression she had of there being something illuminating to be said about 'Someone' – something distinct from 'Look at these uses!' or 'Here is one sort of use of signs, and here is another' – is misleading. And so is the idea we may have, in reading Wittgenstein, that it is easy to come up with examples of propositions that are perfectly intelligible although they lack true–false poles – propositions which are not logically true in a sharp sense, propositions which are not mere tautologies. We may think that Wittgenstein's own book provides us with many such examples: when he speaks of what shows itself, we may think that we grasp the kind of content or quasi-content in question. But what does the showing, if anything does, is not a sign, but a sign in use; and to lay out what does the showing by clarifying propositions would destroy the impression we have of 'what is shown' as something to be said or something to be grasped, a kind of content which we naturally represent to ourselves using a 'that' clause. Only by clarifying propositions – that is, making plain what our symbols are – can we come to recognize that the impression we get of reaching beyond what supposedly can be said is itself a misleading impression. I mean the discussion of Anscombe's example to illustrate that point: she thinks that Wittgenstein held that something she takes herself to grasp (and indeed to be able to communicate) is beyond the reach of saying. My claim is that that impression is created by failure to carry through the task of clarifying what she herself is saying; and that, more generally, the impression of having made sense can be created by failure to clarify what we take ourselves to be talking about and what we take ourselves to be saying about it. This is meant to apply also to the impression we may have that the propositions of the *Tractatus* are intelligible propositions. There is no once-for-all demonstration in the *Tractatus* that propositions of such-and-such sorts are nonsensical: the task of clarifying propositions is a one-by-one task.

Only the activity of philosophical clarification, or of attempting philosophical clarification, can reveal whether, in a particular case, there is or isn't something that we mean.

Notes

1 See 'The Reality of the Past' in G.E.M. Anscombe, *Metaphysics and the Philosophy of Mind* (vol. II of Anscombe, *Collected Philosophical Papers*) (Oxford: Blackwell, 1981): pp. 103–19, especially pp. 112–16, and Anscombe, *Intention* (Oxford: Blackwell, 1963), § 18.
2 G.E.M Anscombe, *An Introduction to Wittgenstein's Tractatus*, 2nd edn (London: Hutchinson, 1963), pp. 85–86.
3 John Koethe, 'On the "Resolute" Reading of the *Tractatus*', *Philosophical Investigations*, **26**, 2003, pp. 187–204, at p. 200.
4 For Flew's discussion, see A.G.N. Flew, ed., *Logic and Language*, first series (Oxford: Blackwell, 1951), pp. 7–8.
5 Ludwig Wittgenstein, *Tractatus Logico-Philosophicus*, trans. D.F. Pears and B.F. McGuinness (London: Routledge, 1961). Hereafter referred to in the text as TLP.
6 This is clear in the TLP 5.1s and 6.12s; see also 'Notes Dictated to G.E. Moore in Norway' in Ludwig Wittgenstein, *Notebooks, 1914–1916*, ed. G.H. von Wright and G.E.M. Anscombe, trans. G.E.M. Anscombe (Oxford: Blackwell, 1961).
7 Ludwig Wittgenstein, *Notebooks, 1914–1916*, *op. cit.*
8 For further discussion of the issues in this paragraph, and more generally of the issues in this essay, see Michael Kremer, 'Mathematics and Meaning in the *Tractatus*', *Philosophical Investigations*, **25**, 2002, pp. 272–303.
9 *Wittgenstein's Lectures on the Foundations of Mathematics, Cambridge, 1939*, ed. C. Diamond (Chicago: University of Chicago Press, 1989), p. 278.
10 For a related view, see Ludwig Wittgenstein, *Zettel*, ed. G.E.M. Anscombe and G.H. von Wright, trans. G.E.M. Anscombe (Oxford: Basil Blackwell, 1967), § 321: 'When a rule concerning a word in a proposition is added to the proposition, the sense of the proposition does not change.'
11 See Michael Kremer's 'Mathematics and Meaning in the *Tractatus*', *op. cit.*, n. 8, on the matter of how different tautologies may show different things.
12 Ludwig Wittgenstein, *Letters to C.K. Ogden*, ed. G.H. von Wright (Oxford: Blackwell, 1973), p. 49. I have spelt out abbreviated words and followed the editor's indicated corrections of punctuation.
13 Flew, *Logic and Language*, *op. cit.*, n. 4, p. 8.
14 See Martin Gustafsson, 'Nonsense and Philosophical Method', forthcoming in S. Pihlström, ed., *Wittgenstein and the Method of Philosophy*, for a discussion of these issues in relation to Wittgenstein's later thought. See also the works of Anscombe referred to in n. 1 for Anscombe's own treatment of the idea of a 'sense that is nonsense' in connection with Wittgenstein's later thought.
15 On these issues, see Michael Kremer, 'Contextualism and Holism in the Early Wittgenstein: From *Prototractatus* to *Tractatus*', *Philosophical Topics*, **25**, pp. 87–120, at p. 98. Kremer lays out the kind of inferential behaviour of propositions in virtue of which the sign '*a*' as used in those propositions is the name of a simple object. Here 'name of a simple object' is a label for the kind of symbol, the revealing sign for which is a variable.
16 See, on these issues, Michael Kremer, 'The Purpose of *Tractarian* Nonsense', *Noûs*, **35**, 2001, pp. 39–73.

- Tractarian use of tautologies for clarificat. purposes exemplified by ref to...
- Translat. of 4.112 (161)
- An example of piecemeal method

Chapter 8

Why Worry about the *Tractatus*?

James Conant

> In order to understand Mr. Wittgenstein's book, it is necessary to realize what is the problem with which he is concerned.[1]

Why worry about Wittgenstein's *Tractatus*? Did not Wittgenstein himself come to think it was largely a mistaken work? Is not Wittgenstein's important work his later work? And does not his later work consist in a *rejection* of his earlier views? So does not the interest of the *Tractatus* mostly lie in its capacity to furnish a particularly vivid exemplar of the sort of philosophy that the mature Wittgenstein was most concerned to reject? So is it not true that the only real reason to worry about the *Tractatus* is to become clear about what sort of thing it was that the later Wittgenstein was most *against* in philosophy? Is the interest of the book therefore not largely exhausted by its capacity to show us what the later Wittgenstein did *not* think?

Much of the secondary literature on Wittgenstein's *Tractatus*, either implicitly or explicitly, answers these questions largely in the affirmative. The aim of this paper is to suggest that the manner in which it has done so has done much to obstruct the possibility of an understanding of Wittgenstein's philosophy – both early and late. The aim is not to suggest that these questions should be answered instead in the negative, but rather to furnish a prolegomenon to the possibility of a proper understanding of what – and how much – ought to be affirmed in answering them in the affirmative.

As the present volume makes evident, there is currently a debate underway about how to read (and how not to read) Wittgenstein's *Tractatus*. This paper will not attempt a direct contribution to that debate,[2] it will attempt instead to bring out some of what might be at stake in that debate. It is natural to think that all that ought to be at stake is a fairly parochial question concerning the proper interpretation of Wittgenstein's work during a single, relatively early phase of his philosophical development. Thus it is natural to conclude that, whatever differences may divide the parties to this debate concerning how to read the *Tractatus*, nonetheless, *au fond* these interpreters of Wittgenstein may be in broad agreement about how to read most of the rest of Wittgenstein's work – or, at least, whatever their disagreements may be about the early work, they are ones that can be independently adjudicated, without substantial cost to anyone's prior commitments concerning how one ought to read the later work. One burden of this paper is to suggest that this is mistaken – that issues parallel to those which arise in the interpretation of the *Tractatus* arise in connection with the interpretation of Wittgenstein's later work as well.

Even though, as stated above, much of the secondary literature on Wittgenstein's *Tractatus* either implicitly or explicitly answers the questions with which this paper begins largely in the affirmative, nonetheless it is customary for it also to acknowledge that there are important *continuities* in Wittgenstein's philosophy. So a different way of putting the question of this paper is as follows: how are these continuities to be conceived? How are we to hold together, in a single unitary account of the development of Wittgenstein's thought, that which he was most concerned to root out and reject in his early philosophy with that in it which he was most concerned to retain and refine?

This is therefore a paper about how to begin to think about the possible continuities and discontinuities in Wittgenstein's philosophy – and hence about where many commentators on Wittgenstein's work think the continuities are to be sought and about where else one might look for them. In what follows, I will attempt to lay bare some of the central exegetical assumptions common to certain standard readings of Wittgenstein's writings at three different stages of his career: the *Tractatus*, the private language sections of *Philosophical Investigations*, and *On Certainty*. This exercise requires that I settle for descriptions of the relevant readings of these works that operate at a fairly severe level of abstraction. I shall abstract, as far as possible, from matters about which the relevant community of readers disagree in order to bring out their fundamental points of agreement. I shall also abstract, as far as possible, from the differences in the doctrines attributed to Wittgenstein at each of these three stages of his development in order to bring out the fundamental continuity in his philosophy, as that continuity emerges on standard readings of his work. In the latter part of the paper, I will indicate why there is reason to think that the philosophical assumptions thus standardly attributed to Wittgenstein are already under indictment in the *Tractatus*, and why there is also reason to think that this indictment is still in force in *Philosophical Investigations* and in *On Certainty*. I hope thereby to offer a brief overview of an alternative picture of where the continuity in Wittgenstein's philosophy might be thought to lie – one which locates it in the opposite place from where we have been taught to look for it by much of the commentary on Wittgenstein's work

A passage from Gordon Baker and Peter Hacker will suffice to give a brief indication of the sort of account of the continuity in Wittgenstein's philosophy that I wish to oppose.[3] The following passage offers a fairly standard story of how an appeal to rules of logical syntax in the *Tractatus* gives way in the later Wittgenstein to an appeal to the rules of grammar:

> Wittgenstein had, in the *Tractatus*, seen that philosophical or conceptual investigation moves in the domain of rules. An important point of continuity was the insight that philosophy is not concerned with what is true and what is false, but rather with what makes sense and what traverses the bounds of sense.... [W]hat he called 'rules of

> grammar' ... are the direct descendants of the 'rules of logical syntax' of the *Tractatus*. Like rules of logical syntax, rules of grammar determine the bounds of sense. They distinguish sense from nonsense.... Grammar, as Wittgenstein understood the term, is the account book of language. Its rules determine the limits of sense, and by carefully scrutinizing them the philosopher may determine at what point he has drawn an overdraft on Reason, *violated the rules for the use of an expression*, and so, in subtle and not readily identifiable ways, traversed the bounds of sense.[4]

I agree with Baker and Hacker that the later conception of grammar is the heir of the earlier conception of logical syntax. I also agree with their proposal concerning the general region of Wittgenstein's thought in which one ought to look for significant continuity: namely, in his conceptions of grammar (or logic) and failures thereof (that is, nonsense) and in his concomitant conception of philosophical method. But I will want to disagree with their particular characterizations of these conceptions – for example, as turning on an aspiration to formulate rules that will 'determine the limits of sense' and thus 'determine at what point' the philosopher has 'traversed the bounds of sense', a point reached when the philosopher 'violate[s] the rules for the use of an expression'. Such characterizations are to be found in much of the secondary literature on Wittgenstein's work – early and late. The question is: how faithful are they to Wittgenstein's own conception of what it is that he was trying to accomplish in philosophy?

Pseudo-Tractarianism

I begin with a brief sketch of the standard reading of the *Tractatus*. It turns on a certain way of understanding the distinction between that which can be *said* in language and that which can only be *shown*. The 'logical form' which a proposition shares with the bit of reality that it depicts cannot itself be made the subject of depiction; rather, it is *shown* in the proposition which possesses it. On the standard way of understanding this, the 'logical form' is conceived of as a *something* – a something which stands in a certain sort of relation to a proposition. So it comes to look as if there are *facts* about what propositions depict: it is a *fact* that this proposition has that logical form.

Having said this much, already at this incipient stage in their exposition, most proponents of the standard reading of the *Tractatus* will begin to backpedal furiously. They will say: 'Well, actually, it is not a fact, but rather a "fact" that this proposition has that logical form.' Some of them think it helps to call what is at issue here a 'fact', placing the word 'fact' in quotes to mark the difference between such 'facts' and ordinary garden-variety facts. Facts are what can be spoken of, what can be depicted by meaningful propositions. What is at issue here is not that sort of fact, but rather something much deeper. Something? Well, not some *thing*. It is something much deeper than a fact or thing. It is like a fact, in that we can, in our thought about it, get it right or wrong; but it lies at too deep a level – deeper than any ordinary fact – to be a mere fact. Let's call those matters which our thought aims to get right, when it seeks to grasp the structure of this deeper domain, *quasi-facts*. Our language cannot depict quasi-facts, we are told, because quasi-facts have to do with what language itself can or cannot depict. That our language cannot depict such quasi-facts comes into view as a limitation on our language. It might be taken to be a merely contingent limitation – one which might be overcome by rising, say, to the level of a meta-language, as Russell suggested in the Introduction to the *Tractatus*. According to the standard reading, the Tractarian response to such a Russellian suggestion is to insist that the limitation in question is a necessary one – to insist that in attempting to frame propositions about 'logical form' we are attempting to state the essential presuppositions of any meaningful language whatsoever. To treat this as a *fact* that could be represented in some other language is to suppose that there could be a language which could accommodate the possibility of that alleged fact's *not* obtaining; and such a language would be an 'illogical language'. And the whole point of the *Tractatus* is, according to the standard reading, to try to show that such a language is something that cannot be. And it is not just something that for some contingent reason cannot be. The non-contingent character of the sort of impossibility here at issue is made manifest to us through our realizing that when we try to speak of what it is that we imagine we might be able to say in a such language (one in which it would be possible to speak of quasi-facts), we end up speaking nonsense. We come to see that any attempt to put into words what it is that we would want to say, when attempting to speak of the essential presuppositions of any meaningful language, must necessarily violate the conditions that render meaningful discourse possible.

If the *Tractatus* is read in this manner, as it standardly is, then it seems to be open to a devastating charge: the procedure of the book as a whole, insofar as it allows itself to talk of how logical form is 'shown', presupposes the possibility of just the sort of language that it seeks to show is impossible. And, indeed, such a charge is difficult to forestall as soon as the difficulty posed by the following question comes into focus: is the 'thought' that 'an illogical language is impossible' itself a thought or not? Is it something thinkable? On the standard reading of the *Tractatus*, Wittgenstein not only bites the bullet here, but seeks to deal with the threat by chewing the bullet up and swallowing it. On the most influential line of interpretation, Wittgenstein's 'solution' to this problem is to relocate it to the realm of the unsayable. So the book is read as attempting to hint at what it cannot say: that there is a realm of ineffable fact-like quasi-truths, such as the fact-like quasi-truth that language and world share a common logical form. On this view, the propositions of the *Tractatus*, while 'strictly speaking' nonsense, serve a useful purpose by directing our attention to the ineffable features of reality and language that undergird all meaningful discourse. In this way, they enable us to 'see the world aright'.

The difficulty is thus transformed from one concerning what we can think and say into one concerning whether certain sorts of ineffable content can be 'shown' and 'grasped'. Usually, the appearance that the difficulty has been surmounted is generated by introducing some device of equivocation. We have already seen one such device: the device of putting expressions in quotation marks ('fact', 'truth', 'grasp' and so on), so that one can deploy a concept and at the same time deny that one is deploying the concept in question, without ever having to make it clear what concept it is that one is thereby deploying, if not the one which figures in quotes. Another favourite device is the handy locution 'strictly speaking'. If the elucidatory propositions of the *Tractatus* are *nonsense*, then how can we understand them? Well, they are, strictly speaking, nonsense (for, after all, the *Tractatus* does say that they are nonsense), but they are not mere nonsense; they are a form of deep nonsense by means of which a special sort of insight can be conveyed. Is what such a proposition conveys a *thought*? Well, actually, it is not, strictly speaking, a thought, (the *Tractatus*, after all, says 'The thought is the significant proposition' (TLP, 4)). But it is like a thought in that one person can communicate it to another and we can grasp what is thus meant. In what sense do we 'grasp' what is 'meant'? Are we able to *understand* that which such a proposition is trying (if failing) to say? Well, strictly speaking, we do not understand it. (The *Tractatus*, after all, says to understand a proposition is to 'know what is the case, if it is true' (TLP, 4.024).). But it is like understanding a proposition, except that in this case there is nothing that is the case for us to understand. Does the *Tractatus* itself distinguish between mere nonsense (that has no meaning) and deep nonsense (whose intended meaning we can grasp)? Well, perhaps not in so many words. But the distinction is required, we are told by commentators, in order to make sense of what the *Tractatus* does want to say, if only it could. Does the *Tractatus* itself distinguish between mere thoughts (that are the contents of significant propositions) and deep 'thoughts' (that outrun the limits of language)? Well, not in so many words. But the distinction is required, we are told, in order to make sense of what the *Tractatus* does want to say, if only it could. And thus it comes to pass that Wittgenstein's work seems to stand in desperate need of the assistance of his commentators in order for it even to appear to be able to stand on its own feet.

I will henceforth refer to readings of the *Tractatus* that dither in such a fashion – both in their treatment of the philosophical problems themselves and in their willingness to credit the *Tractatus* with having thought its own treatment of these problems all the way through – as *irresolute readings* of the *Tractatus.*[5] As I said before, it is not my aim in this paper to argue directly for or against such readings. My aim here is simply to bring out how these readings align with certain alternatives for reading Wittgenstein's later work. But since I do not think such irresolute readings do justice to Wittgenstein's early work, I will allow myself to say that parallel readings of his later work ascribe *pseudo-Tractarian* doctrines to the later Wittgenstein.

Pseudo-Tractarian Readings of the So-called 'Private Language Argument'

On the standard reading of the so-called 'private language argument', Wittgenstein's aim is to show: (1) that the possibility of language-use rests on certain conditions; (2) that a private language does not satisfy these conditions; and therefore (3) that such a language is impossible. If there is an argument of this form to be found in the text, it is buried underneath quite a few distractions and digressions. But the sheer quantity of secondary literature that has been devoted to uncovering it testifies to the extent to which it has become a near-universal article of faith among Wittgenstein scholars that an argument of this form is somewhere there in the text waiting to be excavated. There are a great many differences of interpretation about the details of the supposed argument: disagreements about what the relevant conditions are, about why a private language falls short of these conditions, and hence about why such a language is impossible. But most of these differences of interpretation will not matter for my purposes. What I want to bring out is what almost all such interpretations have in common with one another. They all take Wittgenstein to be trying to show us that there is something that cannot be – a private language. And the reason that it cannot be rests upon a conception of the necessary conditions of the possibility of any language and what it would be to transgress against these.

The 'essential preconditions of our language' now at issue (that is, now that we have turned to the work of the *later* Wittgenstein), tend to be called grammatical truths. Grammatical propositions are taken to be propositions that say something true, and some of the standard ways of evading and confronting the issue of what sort of truth attaches to them ought to remind one of the standard ways of understanding and evading the issue of what sort of pseudo-Tractarian quasi-fact it is that a particular proposition has this, rather than that, sort of logical form. So, now in the interpretation of Wittgenstein's later philosophy, again, a discomfort arises in connection with how we should conceive of the nature of the *something* which a certain sort of proposition, which is not an ordinary factual proposition, asserts to be the case. And here, as before, there is a great deal of wavering in the commentary when it comes to facing this central issue. But, because we now at least know that we are no longer dealing with the Wittgenstein of the *Tractatus*, new scruples come into play about what we are allowed to say about the matter. (Deference to these scruples is often mistaken for philosophical progress.) Again, at this early point in the game, many proponents of the standard reading of a later Wittgenstein backpedal; only now most of them think they have to backpedal in (what at least sounds like) a different direction. They no longer want to retreat from the asserting of philosophical truths to a pseudo-Tractarian showing of them. Yet, in grasping that there cannot be a private language, they still want there to be something that we grasp. We are told that what is at issue here is a 'grammatical truth', rather than an 'empirical truth'; but we are also reassured that such 'truths' are much less metaphysical than pseudo-Tractarian truths about the logical structure of language and reality. On most readings, the quasi-fact that our ordinary factual discourse cannot lay hold of this sort of 'truth' again comes into view as a kind of limitation on our language; only now it is sometimes taken to be a merely contingent limitation – one which might be overcome by switching to a different language. According to such readers of the later Wittgenstein (who have no counterpart among scholars of the *Tractatus*), a fundamental change in the grammatical structure of our language or in our form of life could indeed furnish us with linguistic/conceptual resources able to accommodate the sorts of fact that the present grammar of our language forbids our being able to countenance. (These readers of the later Wittgenstein, like Russell in his Introduction to the *Tractatus*, take the necessities of logic and grammar that Wittgenstein encounters in his philosophizing to be ones that we are able to transcend simply by enriching or exchanging our linguistic resources, and they therefore – though they may be reluctant to acknowledge the point – take the later Wittgenstein to have made progress over his early self through having given way to Russell on this basic point, albeit while differing with Russell over the details concerning the sort of linguistic enrichment required.)

According to other readers of the later Wittgenstein, the proper response to this anthropologistic variant on Russell's suggestion (that the solution to the problems that trouble Wittgenstein is to be found simply through recourse to the vantage point afforded on our own language by another language) is to insist that the 'truths' disclosed by grammatical investigation are indeed necessary ones – to insist that in, for example, attempting to frame 'truths' about the public nature of language-use, we are attempting to state presuppositions of any meaningful language whatsoever. If the 'truths' disclosed by grammatical investigations were mere truths about us (and how we cannot help but think) or about our language (and the way it forces us to carve up the world) or about our culture (and the forms of thinking that it traps us into), then Wittgenstein's view would collapse into some form of psychologism, sociological relativism, or linguistic idealism; and this, these other readers of the later Wittgenstein insist, is just want Wittgenstein most sought to avoid. To treat that which grammatical remarks bring to light – for example, the public nature of language use – as a mere *fact* about *our* language, as these first readers of Wittgenstein do, is to suppose that there could be a language which could accommodate the possibility of this fact's *not* obtaining. But the very notion of such a language makes no sense. To think that there could be such a language is to run foul of that which is disclosed by grammatical investigations regarding such questions as how ostensive definition works, the role of training in language acquisition, and so on and so forth. Thus the whole point of the private language argument, according to this variant of the standard reading, is to try to show that such a language – a private language – is something that cannot be, precisely because it runs foul of what the *Philosophical Investigations* teaches us must be the case in order for any language so much as to be possible. So the target now is no longer the idea of an illogical language, but that of a private language. However, the general shape of the conclusion is the same: a private language is something that cannot be, and not just for some contingent reason. The non-contingent character of this impossibility (and others like it) becomes manifest to us through our realizing that when we try to speak of that which we imagine might possibly be, when we imagine that there could be a private language, we end up speaking nonsense; and we end up speaking nonsense here because we end up violating the conditions of the possibility of meaningful discourse.

If the private language argument is read as proceeding in such a fashion, it, too, would seem to be open to some version of the devastating charge to which the *Tractatus* was formerly alleged to be vulnerable. The argument seeks to show that the very idea of a private language is inherently nonsensical and thus not a possible topic of discourse; but, insofar as talk of 'a private language' is employed with the aim of advancing an argument against the possibility of such a language, the argument would appear to presuppose the possibility of a language in which it is possible to speak of and frame thoughts about a private language – thoughts such as the thought 'a private language is impossible'. Is the 'thought' that 'a private language is impossible' a thought or not? Is it something thinkable? The very structure of such an argument – one that aims to show that the very idea of a private language is one that cannot make sense – seems to presuppose the intelligibility of that which it seeks to show is unintelligible.

One way around this problem is, at this point, to back off from the claim that what grammatical remarks seek to articulate has anything specially to do with the possibilities for making sense (as opposed to speaking truly rather than falsely), and thus to back off from the claim that the aim is to show the philosopher that he lapses into nonsense when he entertains the idea of a private language. One could back all the way off from this claim and claim instead that it is simply an empirical truth that there can never be a private language (so that 'nonsense' is taken to be just a colourful way of saying 'surely false'). A less drastic and more common strategy is to back only halfway off from what seems to be Wittgenstein's own thought about the matter and to try to domesticate Wittgenstein's argument by turning it into one that aims to show that the idea of a private language involves some sort of contradiction in terms (so that 'nonsense' is taken to be just a colourful way of saying 'logically or conceptually contradictory'). Either of these ways of domesticating Wittgenstein's philosophical concerns begins to come free of Wittgenstein's text.

There are commentators who are alert to this, and will simply declare, at this point, that much of what Wittgenstein says – about grammar, about the nature of his method in philosophy, about his wanting to avoid theses, and about his aim being one of taking his interlocutor from latent to patent nonsense and so forth – is quite unfortunate and best disregarded, if one wants to extract those bits which are philosophically useful from his writings. I will have nothing to say to such commentators in this paper, other than to suggest that there is something to be gained by keeping an open mind about whether it is possible that Wittgenstein may have known what he was doing, even if it is not what they themselves want to do in philosophy. This is not to say that there is anything wrong with reading a philosopher simply in order to see what useful ideas you can find him in and not worrying about whether he would approve of the use that you proceed to make of his ideas yourself. (Any philosopher worth his salt often approaches the work of other philosophers in this spirit. Wittgenstein himself certainly often did.) But maintaining the value of such an approach is perfectly consistent with thinking that there is value to approaching, on a different occasion, the work of that same philosopher in a very different spirit – especially if his philosophical approach and sensibility are utterly alien to your own – namely, in a fashion that seeks to understand how he could have ever thought that the bulk of what he wrote hung together as a coherent whole. For it is only by reading such a philosopher in such a manner that one is likely ever to discover the possibility of conceptions of philosophy radically different from one's own. The possibility of such an encounter depends on having an open mind about whether there might be understandings of the nature of philosophical difficulty and the possibility of philosophical progress utterly alien to one's own and yet worth spending one's time to discover. To push to one side, in one's reading of Wittgenstein, everything to do with his understanding of the peculiar nature of the difficulty of philosophy and the character of that difficulty being tied to the question of what is to make sense (and what it is to fail to make sense) is to deprive oneself of the opportunity for such an encounter.

Most commentators on Wittgenstein do continue to want to try to take seriously the idea that, for example, in the private language sections of the *Investigations*, Wittgenstein's aim is to show the interlocutor that he somehow lapses into nonsense. In so far as they seek to read these sections as advancing an argument to the conclusion that what the interlocutor wants to say is nonsense, these commentators are faced with the problem mentioned before: the very structure of such an argument – one that aims to show that the very idea of a private language is one that cannot make sense – seems to presuppose the intelligibility of that which it seeks to show is unintelligible. Some of the commentary, more or less explicitly, tries to get around this problem by exploring the following suggestion: perhaps the appearance of an argument here is a ladder which we are supposed to climb up and then throw away. There are more and less resolute ways to construe such a suggestion. To be a proponent of some version of what I am here calling the standard reading of the later Wittgenstein is to opt for an irresolute construal of the suggestion. In practice this often means throwing away what was supposed to be, according to many proponents of the standard reading of his later work, a fundamental milestone in the progress of Wittgenstein's thought. The relatively uncontentious way to put the point is as follows: it turns out that the later Wittgenstein did not abandon the distinction between saying and showing after all! The more contentious way to put it is as follows: it turns out that the later Wittgenstein was still trying to whistle it! However you put it, on this variant of the standard reading of the private language argument, Wittgenstein's strategy is not to give us an argument to the conclusion that a private language is impossible – for talk of a private language is nonsense, and a claim to the effect that what a bit of nonsense asserts is impossible is itself just a further bit of nonsense. So, according to the reading I am now considering, Wittgenstein's strategy in the private language argument is rather to be seen as seeking to show us something which cannot be asserted in a factual proposition or in any other kind of proposition, not even a grammatical proposition. The thing he wants to show us, if we try to put it into words, will, admittedly, end up sounding something like the following remark: 'A private language is impossible.' But, according to this unabashedly pseudo-Tractarian reading, what this remark seeks to convey is a kind of insight that cannot really be conveyed by an ordinary assertoric proposition. Nevertheless, it is something that can be shown – and that we can come to grasp – by working through Wittgenstein's investigations. A grammatical proposition therefore does not succeed in saying a sort of thing that mere factual propositions cannot say, for what it appears to say simply is not assertable. Rather what it does is merely to summarize in a helpful shorthand that which is shown – and can only be grasped by someone who has successfully worked through the exercise of following all the way through what is shown – in a grammatical investigation. If one construes what is thus summarized as a sort of truth that cannot be stated and yet which can be conveyed, then one remains within the ambit of a pseudo-Tractarian interpretation of Wittgenstein's later philosophy.

Proponents of the standard reading of the so-called private language are therefore faced with a dilemma. Either they leave themselves defenceless against a version of the devastating charge or they avoid it by employing what is essentially a pseudo-Tractarian gambit. (Or, more typically they hover between these two options, without ever even noticing that they so hover.) The typical proponent of the last of the aforementioned variants of the standard reading of the later Wittgenstein – the one that has a use for a pseudo-Tractarian construal of the idea that Wittgenstein holds on to the distinction between saying and showing in his later work – will, in my experience, immediately bristle if one suggests that his 'solution' to the problem of the devastating charge is to relocate the problem to the realm of the unsayable. Although such a reader will view the private language sections as attempting to reveal something that cannot simply be asserted, unlike the corresponding reader of the *Tractatus*, he will not want to saddle the later Wittgenstein with the doctrine that there is a realm of ineffable fact-like quasi-truths called grammatical truths. Yet this same reader of Wittgenstein, his distaste for ineffability theses notwithstanding, does want to say that the body of quasi-truths displayed in Wittgenstein's grammatical remarks possess the following features: they cannot be expressed in factual language and can only be shown through a grammatical investigation. Moreover, generally such a reader will not want to call that which these grammatical remarks allow us to grasp facts, or even 'facts'. Therefore, on this reading of the later Wittgenstein, the grammatical propositions of the *Philosophical Investigations* are at least tacitly held to possess all of the following pseudo-Tractarian characteristics: they seek to give voice to truths which are not garden-variety truths; what thus comes into view are not facts; these 'truths' are 'strictly speaking' unassertable; the propositions that attempt to state them nonetheless serve a useful purpose; this purpose is achieved by directing our attention to features of language and/or reality that undergird the possibility of meaningful discourse and, in this way, they enable us to see matters aright.

Such readers of the later Wittgenstein will do their best to describe their own reading of Wittgenstein in ways that disguise the extent of its pseudo-Tractarian character. They do not want their reading of the later Wittgenstein to sound too much like their reading of the *Tractatus*. Some of these readers of Wittgenstein will admit to the existence of a certain degree of parallelism. They might agree that, for the later Wittgenstein as for the early Wittgenstein, philosophy does not consist in a body of doctrine but an activity, and that the purpose of philosophy is not to arrive at a set of philosophical propositions but to arrive at a certain way of viewing language and/or world. But they will do their best to emphasize those features of the later Wittgenstein's philosophical practice which highlight the differences between the later Wittgenstein (on their reading of him) and the early Wittgenstein (on their reading of him). They will emphasize the role of examples and the attention to the use of language in the work of the later Wittgenstein. They will emphasize the insistence on the existence of a multiplicity of grammars and the importance of the role played by an attention to the motley of kinds of language-game in the later work. They will emphasize that language-games are instances of language-use interwoven with human activity and that everything now turns on an appreciation of just how deeply context-dependent each such instance is. And so on. And they will attempt to draw upon all of these differences in the descriptions they offer of what it is that is shown by the later Wittgenstein and how it is shown, so that it sounds nothing like their descriptions of the activity by means of which the *Tractatus* allows us to grasp that which is shown but cannot be said. But the true measure of how pseudo-Tractarian a reading of the later Wittgenstein is is not to be gauged by the vocabulary in which it is couched. You can replace talk of 'logical syntax' with 'grammar', 'language' with 'language-games', 'application' with 'use', 'elucidation' with 'investigation' and so on and still end up attributing a doctrine to the later Wittgenstein that is mired in an essentially pseudo-Tractarian problematic.

Pseudo-Tractarian Readings of *On Certainty*

Standard readings of the argument of Wittgenstein's very last work, *On Certainty*, tend to be equally pseudo-Tractarian. The argument here, again, turns on a distinction between two sorts of proposition – in this case, propositions that are situated within the framework of our practices of making and accepting knowledge-claims and propositions that seek to articulate constitutive features of that framework. Thus what sceptical reflections and Moore-style responses to scepticism reveal, according to the standard reading of Wittgenstein's last work, is that there are these two types of proposition or judgement: there are those that are part of the framework and those that are not. Let us call the former 'framework propositions'. (They are also sometimes called 'Moore-type propositions' or 'hinge propositions'.) Leaving aside differences in terminology and nuances of doctrine, the same fundamental difficulties tend to accumulate around this privileged category of propositions (on the standard reading of what they are supposed to be) that we earlier saw attend Tractarian propositions about logical form, on the one hand, and *Investigations*-style grammatical propositions, on the other (on the standard readings of what each of these is supposed to be). In each of these three cases, we need a distinction that allows us to articulate a difference between two sorts of proposition: those propositions which say that which can simply be said without further ado and those propositions which articulate a limit which marks off sense from nonsense, but which are unable to pull off this trick of articulation without considerable further ado.[6] According to the standard reading of this matter, framework propositions, in ordinary contexts, are not subject to confirmation or disconfirmation. Rather, they form the fixed points around which all inquiry revolves – they constitute the framework which makes all inquiry possible – and, as such (as long as they are caught up in playing this framework-constituting role), they themselves cannot come into question. Any of the following, we are told by proponents of this reading, if uttered on an appropriately inappropriate occasion, would constitute an example of a framework proposition: 'The world existed a long time before my birth'; 'Things do not go in and out of existence'; 'Everyone has parents'; 'My name is James Conant'; 'This is my hand'; 'I am a human being'; 'I am presently reading a paper to an audience'; 'I am here'. There are differences among proponents of the standard reading about what feature, or features, mark propositions such as these out, on certain occasions of use, as framework propositions. Sometimes the crucial feature seems simply to be that, on a certain occasion of use, what is said is 'flamingly obvious'; sometimes it seems to have more to do with the authority which the proposition or the speaker of the proposition possesses; sometimes it seems to have more to do with the manner in which such a proposition must be taken for granted in order for inquiry to proceed. According to most accounts, which of these features is relevant itself depends on the particular case of the framework proposition in question. Such differences in interpretation, again, will not matter for my purposes here.

What does matter is that, on most readings of *On Certainty*, framework propositions are taken to be propositions that attempt to say something true; and, once again, the standard ways of evading and confronting the issue of what sort of truth attaches to them ought to remind one of standard ways of dealing with, and evading, the issue of what sort of pseudo-Tractarian 'fact' it is that language has this, rather than that, sort of logical form.[7] Here, again, a discomfort arises in connection with how we should conceive of the nature of the *something* which this sort of proposition, which is not an ordinary factual proposition, asserts to be the case. Again, to the ears of most commentators on the latest Wittgenstein, it will sound altogether too Tractarian to say anything like that there is a state of affairs which stands in a truth-making relation to a framework proposition, or that what is asserted by a framework proposition is a 'fact' in any ordinary sense of the term. So, at this early point in the game, almost all proponents of the standard reading of the latest Wittgenstein also backpedal, only now most of them think they have to backpedal in (what at least sounds like) yet another direction. They will think that it still helps to insist that the sort of 'truths' articulated by framework propositions are not 'empirical truths' (how could they be if they are neither confirmable nor disconfirmable?), only now they often think that it helps to mark the difference between such 'truths' and ordinary garden-variety truths by striking the note of *activity* – of doing rather than saying, of knowing-how rather than knowing-that.

Under ordinary circumstances, if we say 'I am here', we will not be understood by others; and we will not be understood because it will not be clear what we are *doing* with our words. Now that certainly seems right. But most readers of the latest Wittgenstein want to go on to suggest something further: that the very nature of our practice renders it impossible to say this sort of thing – it renders it nonsense. Now what is nonsense here? Is it the sentence – that is, the form of words – that is nonsense? How can that be? I can, after all, come up with circumstances in which it makes perfect sense to utter this sentence. (I push the button next to your name and speak the following words into the intercom: 'I am here.') If the sentence can make perfect sense then it is not just the sentence *per se* that is nonsense. It must have something to do with the circumstances of use. But still the question remains, when uttered under such circumstances, what is nonsense? Is it what we say that is nonsense? Well, if the account of why it is nonsense is supposed to have to do with how our practice of entering and assessing claims forbids us from saying things that are flamingly obvious, then there must be something which gets said such that it can be claimed that *it* is flamingly obvious. So perhaps what is said makes sense, but the saying of it does not. Perhaps it is the *act* of saying it under certain circumstances that is generative of nonsense.

Does this mean that the epithet 'nonsense' applies only to the speech-act and not at all to the content of what is thereby asserted? In the face of questions such as these, the dithering begins. But despite all the dithering, this much tends to remain fairly clear: there is supposed to be something that we cannot say here that we are nonetheless supposed to come to grasp; and it is in grasping this that we come to appreciate something about the structure of our linguistic practice – about the shape of our framework. So that there is a sort of 'truth' that our ordinary factual discourse cannot lay hold of again comes into view as a kind of limitation on the expressive resources of our linguistic practice. It might be taken to be a merely contingent limitation – one which might be overcome by switching to a different linguistic practice.

Interestingly, now, when we turn to *On Certainty*, the constituency of readers of the later Wittgenstein who favour such a reading swells in number. A change in our everyday practices of claim-making, these readers tell us, could indeed furnish us with linguistic/conceptual resources able to accommodate the sorts of claim that the present structure of our linguistic practice forbids us from ever being able to countenance as candidates for truth or falsity. According to other readers of *On Certainty*, however, the proper response to this reprise of the anthropologistic variant on Russell's suggestion is to insist that the 'truths' disclosed by framework propositions are never susceptible to doubt, since to allow otherwise would be to concede the truth of scepticism. (The sceptic isn't denying after all that, while caught up in our everyday practices, we find certain claims all but impossible to doubt. He is claiming that however unnatural we may find it to do so, these seeming indubitable claims can, when subjected to the appropriate pressures, be shown to be susceptible to sceptical doubt.) According to these other readers, to treat the indubitability of framework propositions as a mere artefact of our practice, as the first sort of readers of Wittgenstein wish to do, is to suppose that there could be a linguistic practice which could accommodate the possibility of these propositions' not obtaining while remaining the sorts of propositions that they are. This would require a practice which would allow us, at one and the same time, to mean what we mean by them and yet to look upon them as possibly false. But such a practice would be an incoherent practice. To think that there could be such a practice is to run foul of Wittgenstein's fundamental insights regarding the manner in which inquiry must be conducted. Such a practice would require that we be able to accept, and at the same time refuse, the fundamental techniques of description and claim-making presupposed by the practice as a whole. And the whole point of Wittgenstein's last work, according to this variant of the standard reading, is to try to show that such a practice – one that could accommodate sceptical doubts regarding the epistemic credentials of its own framework – is something that cannot be. It is not just something that cannot be for some contingent reason. The non-contingent character of this impossibility becomes manifest to us through our realizing that when we try to convert features of our framework of inquiry into topics of inquiry, we end up speaking nonsense; and we end up speaking nonsense because we end up violating the conditions of the possibility of meaningful discourse. Such nonsense arises whenever we try either to question or to claim to know the truth of a framework proposition.

If the argument of Wittgenstein's last book is read as proceeding in such a fashion, it, too, would seem to be open to some version of the devastating charge to which the *Tractatus* was previously held to be vulnerable. If the sceptic's questions and Moore's counterassertions are nonsense, then so, too, should be the framework propositions which seek to articulate and display those 'truths' – or quasi-truths. But how can the conclusion that they are nonsensical be made to cohere with the claim that they are fundamental to our linguistic practice and that that is why they cannot be doubted in the manner of the sceptic nor affirmed in the manner of Moore? Some proponents of this interpretation try to buy room by saying that framework propositions are not nonsensical; they are just not assertable. Yet it is hard to see how something which can never be asserted can be meaningful. And, if what a framework proposition says is meaningless, it is hard to see how it can also be part of a framework. How can that which is asserted by a piece of nonsense be fundamental to anything? Is the thought expressed by a framework proposition a thought or not? Is it something thinkable? Once again, the very structure of the argument attributed here to Wittgenstein seems to presuppose the intelligibility of something which, at a later stage, it is committed to declaring unintelligible. And, here again, it is tempting to think that the following stratagem might buy one philosophical breathing space: the appearance of an argument here is ultimately to be recognized as a ladder which we are supposed to climb up and then throw away. On this variant on the standard reading of Wittgenstein's last work, the strategy is not to show us that what we assert when uttering a framework proposition is true – for any attempt to assert such a proposition is nonsense, and nonsense has no truth-conditions. Wittgenstein's strategy is to show us something about the nature of our linguistic practice that cannot be asserted in a factual proposition or in any other kind of proposition – not even in a framework proposition. Although what these propositions seek to assert can not be asserted, nevertheless, it is something that can be shown – and that we can come to grasp – through an examination of the nature of our lives with language. A framework proposition therefore does not say something that cannot be said in factual language, for what it appears to say simply is not assertable. It merely points to something which can be shown – and can only be grasped by someone who has successfully worked through the exercise of following all the way through what is shown – in the appropriate sort of examination of what it is that we actually *do* with language. Here again, if one construes what is thus pointed to as a sort of truth that cannot be stated and yet which can be conveyed, then one remains within the ambit of a pseudo-Tractarian interpretation of Wittgenstein's later philosophy.

A proponent of this latter sort of reading of the latest Wittgenstein will, again, bristle if one suggests to him that his 'solution' to the problem posed by the devastating charge is simply to relocate the problem to the realm of the unsayable. He does seem to think that framework propositions hint at something they cannot say, but he does not want to saddle the latest Wittgenstein with a last-minute conversion back to the pseudo-Tractarian doctrine that there is a realm of ineffable fact-like framework truths. His distaste for such a doctrine notwithstanding, he will insist that what framework propositions misfiringly attempt to assert are not facts, or even 'facts'. Indeed, what they point to are not even supposed to be candidates for assertion, yet the attempt to assert them still serves a useful purpose by directing our attention to features of linguistic practice and/or reality that undergird our linguistic practice. In this way, they enable us to see matters aright.

And, again, these readers of the latest Wittgenstein will do their best to describe their reading of Wittgenstein in ways that disguise its pseudo-Tractarian character. They will emphasize that we are now no longer talking about the logical structure of language, but rather about the presuppositions of a practice; we are not talking about what we can say in language, but about what we do with language and so on. But, again, do not be fooled: the proper measure of how pseudo-Tractarian a reading of the latest Wittgenstein is is not to be gauged by the vocabulary in which it is couched. It is to be gauged by the structure of the problems it encounters, the nature of the responses these engender, and the attendant forms of evasion and wavering that characterize such responses.

structure of probs.

The Pseudo-Continuity of Wittgenstein's Philosophy

I have sought thus far in this paper only to bring out how deep the parallels run between standard readings of the early, the later, and the latest Wittgenstein. I do not mean hereby to deny that one can discover all sorts of difference within these parallel conceptions attributed to Wittgenstein at these supposedly very different stages of his development. Nonetheless, it is important to come to see how very little philosophical progress Wittgenstein actually succeeds in making with his fundamental problems, if the standard narrative of his intellectual trajectory has any merit. The parallel begins, at each stage, with the idea that Wittgenstein is concerned to show that there is something that cannot be: there cannot be an illogical language; there cannot be a private language; there cannot be a practice of knowing and doubting whose framework judgements are themselves candidates for knowledge or doubt. Or, if one thinks that it helps matters, one can reformulate the putative parallel here in linguistic terms. Thus reformulated, the parallel begins with the idea that, at every stage of his career, Wittgenstein is concerned to show that we run up against a limit of language: we cannot say what can only be shown; I cannot speak about anything so private that others are necessarily incapable of speaking about it; we cannot say we know or doubt the judgements expressed by framework propositions. Or, if it seems insufficiently Wittgensteinian to couch matters in general terms rather than in terms of specific examples of philosophical nonsense, one can reformulate the parallel by displaying particular instances of the particular sort of thing that our logic, or grammar or linguistic practice allegedly debars us from saying: I cannot say this proposition has that logical form; I cannot say 'Only I can know my pain'; I cannot claim under present circumstances to know that I am here. However one formulates the parallel here, in each case the parallel lies in the idea that *there is something we cannot do*. To think that this is the locus of significant continuity is to miss what I take the real continuity in Wittgenstein's philosophy to be – one which is nicely summarized in section 374 of *Philosophical Investigations*: 'The great difficulty here is not to represent the matter as if there were something one *could not* do.'

The idea that Wittgenstein's aim is to show us that there is something we cannot do goes hand-in-hand with the tendency to read Wittgenstein, at each stage of his career, as seeking to adumbrate the conditions of meaningful speech – whether one takes these conditions to reside in the logical structure of language, the rules of grammar or the shape of our linguistic practice. In each case, Wittgenstein's aim is taken to be one of showing the philosopher that he is speaking nonsense because he has violated certain conditions. The integrity of Wittgenstein's method of philosophical criticism thereby comes to apparently require that he first be able to specify what the conditions in question are. This requirement, in turn, gives rise to the appearance that he must be committed to the existence of a set of quasi-truths – whether they be called the principles of logical syntax, the rules of grammar, or the judgments that constitute the framework of our practice – that mark the bounds of sense and reveal the exact point at which the philosopher has strayed beyond them. These quasi-truths must lie outside the realm of merely empirical truths, while lying just inside the limit beyond which the realm of impermissible philosophical nonsense begins.

Once the question arises what sort of truth it is that these quasi-truths can be said to be possess, the waffling and dithering begins. For the account of nonsense that emerges from the specification of such a set of conditions on the possibility of meaningful discourse always requires a distinction between two sorts of proposition: ordinary empirical propositions and logical, grammatical, or framework propositions – it being through the latter that the conditions on the former are to be articulated, and the problem being that the latter always end up looking as if they involve what is, by their own lights, a failure to live up to the conditions of meaningfulness that they themselves seek to articulate. But the full extent of the parallel between readings of the early, the later and the latest Wittgenstein only comes fully into view when one learns to recognize the homology in the steps of the dance that ensues when readers of each these

Wittgensteins attempt to evade the consequences of some version of what I have called the devastating charge. The first sequence of steps begins with an attempt to domesticate the conditions on meaningful speech by construing them as contingent limitations on our capacities for thought and language, and it ends up collapsing into some form of psychologism, sociologism or linguistic idealism. The second sequence of steps begins with an attempt to affirm the non-empirical character of the statements that formulate the conditions of the possibility of meaningful speech, and it ends up leaving it a complete mystery why the putatively less than nonsensical logical, grammatical or framework propositions should be thought to be in any better (logical or grammatical) shape than the propositions they seek to expose as nonsensical. The third sequence of steps begins by biting the bullet and accepting the charge that the propositions that articulate the conditions on meaningful speech are themselves meaningless – like the statements they seek to condemn – but it then tries, by whistling what cannot be said, to find a way for these bits of nonsense to succeed in communicating (that which they cannot say) by some other means. It is when this third sequence of steps reaches its climax that readings of the later Wittgenstein become unmistakeably pseudo-Tractarian. But it is mistake to think that a reader who shrinks from following through on all the steps of the dance, and takes pride in having remained stuck somewhere in the middle of the first or second sequence of steps, has thereby avoided the dialectic of pseudo-Tractarianism. The whole point of the *Tractatus* is that all of its rungs belong to a single ladder. To execute the final sequence of steps in this dance is at least to have climbed all the way up to the penultimate rung of the ladder, and, comparatively speaking, that is a form of progress. But it is not yet to do what the *Tractatus* seeks to enable its reader to do: namely, to throw the whole ladder away.

An Alternative Reading of the *Tractatus*

I have termed the reading of the *Tractatus* briefly sketched at the beginning of this paper pseudo-Tractarian in order to be able to distinguish it from the reading of that work which I myself favour and which I have defended in other writings of mine. I cannot go into the details of that reading here,[8] but will only say enough to indicate why, if one endorses that reading, one will want to conclude that the early Wittgenstein had already set forth, within the pages of the *Tractatus,* a critique of the doctrines attributed to the early, the later and the latest Wittgenstein on the standard readings of each of these three stages of his philosophical development outlined above – a critique that does not attempt to pretend that the devastating charge is anything less than devastating.

The fundamental differences in different readings of the *Tractatus* are usually to be traced to differences in how readers approach the famous penultimate paragraph of the book, where we are told that the author's propositions serve as elucidations by *our* – that is, the reader's – coming to *recognize* them as nonsensical. The question any reading must come to terms with is this: how can the recognition that a proposition is *nonsense* ever elucidate – ever shed light on – anything? This is what the *Tractatus* has to say about what is distinctive about its own conception of nonsense:

> Frege says: Every legitimately constructed proposition must have a sense; and I say: Every possible proposition is legitimately constructed, and if it has no sense this can only be because we have given no *meaning* to some of its constituent parts. (TLP 5.4733)

I would like to draw attention to the word 'only' in this passage. The critical difference between the formulation attributed to Frege and the one which the *Tractatus* endorses is that the former implicitly distinguishes between those propositions that are legitimately constructed and those that are not, while the latter rejects the idea that there is such a thing as a logically illegitimately constructed proposition: 'Every *possible* proposition is legitimately constructed.' The 'only' indicates that there is only one way for a proposition to fail to have a meaning where one might have thought that there were two. What does it mean to reject the idea that there could be logically illegitimately constructed propositions? Or, to put the same question differently, what does the early Wittgenstein mean, when he says (in TLP 5.4732) 'We cannot give a sign the wrong sense'?

I have argued elsewhere – in a paper titled 'The Method of the *Tractatus*'[9] – that Wittgenstein saw a tension in Frege's thought between two different conceptions of nonsense, which I call the *substantial conception* and the *austere conception* respectively. The substantial conception distinguishes between two different kinds of nonsense: mere nonsense and substantial nonsense. Mere nonsense is simply unintelligible – it expresses no thought. Substantial nonsense is composed of intelligible ingredients combined in an illegitimate way – it expresses a logically incoherent thought. According to the substantial conception, these two kinds of nonsense are logically distinct: the former is mere gibberish, whereas the latter involves what commentators on the *Tractatus* are fond of calling a 'violation of logical syntax'.[10] The austere conception, on the other hand, holds that mere nonsense is, from a logical point of view, the only kind of nonsense there is. The *Tractatus* is standardly read as championing the substantial conception. This is, I argue, to mistake the bait for the hook – to mistake the target of the work for its doctrine. On the reading of the *Tractatus* I sketch in the above-mentioned paper, the *Tractatus* is to be seen as resolving the tension in Frege's thought between these two conceptions of nonsense in favour of the austere view.[11] The presence of the word 'only' in the above passage is therefore to be seen as signalling Wittgenstein's rejection of the substantial conception.

On a resolute reading, the *Tractatus* itself does not subscribe to any variant of the substantial conception, and thus the aim of the work is not to show us that certain sequences of words (or particular employments of them) possess an intrinsically flawed sense by persuading us of the truth of some theoretical (though perhaps unsayable) doctrine about where to locate 'the limits of sense'. Rather, the early Wittgenstein seeks to show that any theory which seeks to draw such 'a limit to thinking' commits itself, as he says at the outset of the book, to being 'able to think both sides of the limit' and hence to being 'able to think what cannot be thought'. The Tractarian attack on substantial nonsense – on the idea that we can discern the determinately unthinkable thoughts which certain pieces of nonsense are trying to say – is an attack on the coherence of any project which thus seeks to mark the bounds of sense. The *Tractatus* seeks to bring its reader to the point where he can recognize sentences within the body of the work as nonsensical not by means of an account of the conditions of meaningful speech which legislates certain sentences out of the realm of sense, but rather by bringing more clearly into view for the reader the life with language he already leads – by harnessing the capacities for distinguishing sense from nonsense implicit in the everyday practical mastery of language which the reader already possesses. As the Preface says, 'The limit ... can only be drawn *in language* and what lies on the other side of the limit will be simply nonsense'. The word 'simply' here goes with the 'only' (in TLP 5.4733) to which I drew attention previously. That is, once we go 'outside' of language what we end up with are forms of words which are nonsense in what this work aims to show is the only way anything can be nonsense. The work seeks to do this not by instructing us in how to identify determinate cases of nonsense, but by enabling us to see more clearly what it is we do all the time with language when we succeed in achieving determinate forms of sense and what it is we fall short of doing when we fail to achieve such forms of sense, yet are prey to the illusion that we are making a kind of sense.

The assumption underlying Tractarian elucidation, on this reading of the book, is that the only way to free oneself from such illusions is to enter into them and explore them from the inside. The illusion that the *Tractatus* seeks to explode, above all, is that we can run up against the limits of language. The book starts with a warning about a certain kind of enterprise – one of attempting to draw a limit to thought. In the body of the text, we are offered (what appears to be) a doctrine about 'the limits of thought'. With the aid of this doctrine, we imagine ourselves to be able both to draw these limits and to see beyond them. We imagine ourselves able to do what the Preface warns we will fall into imagining ourselves able to do (once we imagine ourselves able to draw a limit to thought): we imagine ourselves able 'to think both sides of the limit' (and hence 'able to think what cannot be thought').[12] The aim of the work is to show us that beyond 'the limits of language' lies not ineffable truth, but rather (as the Preface cautions) *einfach Unsinn*.[13] At the conclusion of the book, we are told that the author's elucidations have succeeded only if we recognize what we find in the body of the text to be nonsense. He tells us in TLP 6.54 how these sentences serve as elucidations: by enabling us to recognize them *as* nonsense.[14] The sign that we have understood the author of the work is that we can throw away the ladder that we have climbed. That is, we have finished the work, and the work is finished with us, when we are able to *throw* the sentences in the body of the work – sentences about 'the limits of language' and the unsayable things which lie beyond them – *away*.

An Alternative View of the Continuity in Wittgenstein's Philosophy

We are now ready to take up the question where the continuity in Wittgenstein's philosophy will appear to lie to someone who undertakes to read all of his major writings – early and late – in a similarly resolute fashion. Consider the following remark from 1935:

> *Different kinds of nonsense.* Though it is nonsense to say 'I feel his pain', this is different from inserting into an English sentence a meaningless word, say 'abracadabra' (compare Moore last year on 'Scott kept a runcible at Abbotsford') and from saying a string of nonsense words. Every word in this sentence is English, and we shall be inclined to say that the sentence has a meaning. The sentence with the nonsense word or the string of nonsense words can be discarded from our language, but if we discard from our language 'I feel Smith's toothache' that is quite different. The second seems nonsense, we are tempted to say, because of some truth about the nature of things or the nature of the world. We have discovered in some way that pains and personality do not fit together in such a way that I can feel his pain. – The task will be to show that there is in fact no difference between these two cases of nonsense, though there is a psychological distinction, in that we are inclined to say the one and be puzzled by it and not the other. We constantly hover between regarding it as sense and regarding it as nonsense, and hence the trouble arises.[15]

For a resolute reader of Wittgenstein, Wittgenstein's description here of his task – to show that there is in fact no logical difference between these two cases of nonsense – will seem equally accurate as a description of the task of his early and his later work. Certain passages in the later work, however, in which Wittgenstein speaks, for example, of 'excluding certain combinations of words from our language', might seem to contradict this, thus inviting a reading of Wittgenstein along the following lines: certain combinations of words are to be identified as impermissible on the grounds that these combinations violate the principles governing which combinations of words are grammatically well-formed. It is precisely such a reading of his work that a resolute reader of Wittgenstein will hear him seeking to fend off in § 500 of *Philosophical Investigations*: 'When a sentence is called senseless, it is not as it were its sense that is senseless.' A resolute reader of Wittgenstein will take such a passage to mark an important continuity in Wittgenstein's thought: when Wittgenstein remarks in his later writings that a nonsensical word does not have a 'senseless sense' (as for example, in *Philosophical Investigations*, § 500), he is refashioning the Tractarian point that we cannot give a sign 'the wrong sense'.

To see what is at stake here, recall that according to the standard reading of the so-called private language argument, Wittgenstein's aim was to show: (1) that the possibility of language use rests on certain conditions; (2) that a private language does not satisfy those conditions; and therefore (3) that such a language is impossible. Such a reading of the 'private language argument' takes the aim to be to show that the very idea of a private language is the idea of something which we can rule out because of the kind of sense that the locution 'private language' has antecedently been given. It takes it that there is something determinate which the philosopher wants to mean by the locution 'private language' and that *that* is nonsense. There are, of course, all kinds of perfectly determinate things one might mean in speaking of a 'private language'. I might tell you that I just invented a private language and that I am its only speaker, but, if you are nice to me, I might consider teaching it to you. But that is not the kind of thing the philosopher wants to mean by a private language. Why not? Because, we will be told, it is not private enough. Wittgenstein says his topic is 'the language which describes my inner experiences and which only I myself can understand'; and we will be told that the right way to understand what is at issue here is to understand that a 'private language' is a sort of language which it is impossible for anyone but me to understand. Thus there is a right way of understanding what Wittgenstein means by 'private language' and this is the sort of thing that he is concerned to show cannot be. There is therefore supposed to be a determinate something that is the thing that the philosopher does want to mean that Wittgenstein is taken to be interested in singling out in order to show that *that* is something that the philosopher cannot mean because it is nonsense. On any such analysis of what is at issue here, what the philosopher means must be shown to be not mere nonsense but substantial nonsense.

What would a reading of the relevant sections of *Philosophical Investigations* that did not attribute to Wittgenstein a commitment to the substantial conception of nonsense look like? A resolute reader of this stretch of Wittgenstein's later work will want to begin by pointing out that Wittgenstein does not say that there cannot be a private language. The discussion of private language (like so many passages in Wittgenstein's later work) begins with an invitation for us to try to imagine something.[16] The point of the exercise is not to get us to see that there is something determinate to imagine which we are then supposed to see as a sort of thing that cannot be. Rather, the point of the exercise is to get us to see that there is nothing for us to mean by the locution 'private language' that corresponds to what we, under the pressure of certain philosophical perplexities, want to mean by it. There are, of course, various things we can mean by it, but they do not answer to our philosophical desires. And as long as there hovers before us the seeming possibility of something further to mean that does promise to answer to our philosophical desires, the philosopher's treatment of the question of a 'private language' remains unfinished. The point of the exercise of trying to imagine 'a private language' is to work through 'the seeming possibility' here – to try to think it all the way through – until we find it dissolves on us. The transition from latent to patent nonsense is the point of the exercise: the task is to help us see how the seeming possibility (of a 'private language') dissolves under the pressure of an attempt to work out what 'it' (seemingly) requires. What makes a reading of this bit of Wittgenstein's text 'resolute' (in my parlance) is that it follows all the way through on such an exercise to the point where the apparent something dissolves into a nothing. And what makes a reading 'irresolute' (in my parlance) is that it continues to cling to the idea that there is something determinate which the words 'private language', on a philosophical employment of them, are trying, but failing, to mean, and that

Wittgenstein's aim is to show his interlocutor that *that* is a sort of thing that cannot be. To fail to see how different these two different ways of reading Wittgenstein are is, I think, to fail to see what is at stake in his work – what sort of treatment of philosophical problems he seeks to offer – both early and late.

On some readings of the so-called 'private language argument', the nonsensicality of the idea of a private language can be traced to the incompatibility of the concepts 'private' and 'language'. These two concepts, we are told, cannot be made to fit together. On most readings of *On Certainty*, on the other hand, the nonsensicality of uttering a proposition such as 'I am here' under ordinary circumstances is to be traced to a different sort of incompatibility – the incompatibility of the sentence with its context of use. This sentence and this context, we are told, cannot be made to fit together. It is important to see that the progress from one of these accounts of nonsense (in terms of the incompatibility of the components of a proposition) to the other (in terms of the incompatibility of the components of a speech-situation) does not make the sort of difference that some commentators on Wittgenstein's work would like to imagine it does.[17] Each of these accounts holds that there is something determinate that the philosopher wants to mean, if only he could. As long as one traces each of the stages of progress in Wittgenstein's philosophy in terms of such exchanges of one version of the substantial conception of nonsense for another – as almost all of the commentary on Wittgenstein's work does – one has made Wittgenstein's philosophy, in all of its stages, vulnerable to the sort of criticism that he sought already in the *Tractatus* to advance against the views of Frege and Russell.

Consider what Wittgenstein says in *On Certainty*, § 348:

> [T]he words 'I am here' have a meaning only in certain contexts, and not when I say them to someone who is sitting in front of me and sees me clearly – and not because they are superfluous, but because their meaning is not *determined* by the situation, yet stands in need of such determination.

What Wittgenstein says here is not that it is clear what determinate thing it is that the sentence 'I am here' means, but, because of the unsuitable character of its context of use, it cannot be said. Rather what he says is that the meaning of these words is not determined by the situation – their meaning still stands in need of further determination. Over and over again, in *On Certainty*, Wittgenstein tries to get us to see that it is only through its being caught up in circumstances of use that a sentence has sense. This is why he says in *On Certainty*, § 348 that 'the words 'I am here' have a meaning *only* in certain 'contexts' – that is, it is a mistake to think that the words themselves possess a meaning apart from their capacity to mean what they do in the various contexts of use in which they can be put to work to say something. All of the differences in philosophical approach between the two works notwithstanding, the 'only' here in *On Certainty* is making a point that parallels the point of the 'only' in 5.4733 of the *Tractatus*. The point of the 'only' is that there is not some other way for the words 'I am here' to achieve a determinate sense apart from having one conferred upon them through their employment in a context of use; and hence there is not some other way for these words to fail to make sense apart from such a failure to confer determinate sense upon them. And this means that there is not some way for these words to fail to make sense when imported into a particular context of use because of the fully determinate sense they already have, prior to and independent of their occurrence in contexts of significant use. Here, again, the 'only' signals Wittgenstein's commitment to the austere conception of nonsense: the only way for a form of words to be nonsensical is for us to fail to mean something determinate by them.

Wittgenstein sought in his early work to show us how we are able to fail to confer a determinate method of symbolizing upon a sign while imagining that we have already done so. His early method of philosophy tried to get us to see this by having us see that some putative thought (that we imagined ourselves able to lay hold of by calling upon a particular form of words) could not be translated into (what the *Tractatus* calls) a proper logical syntax – an ideography specifically designed to perspicuously display the logical structure of any proposition expressed in it. The point of employing such a notation was not to demonstrate to the philosopher that he had violated a principle of logical syntax. Rather, faced with a perspicuous representation in a proper logical symbolism of the possibilities available to him for meaning his words, the philosopher was to discover for himself that he had unwittingly hovered between alternative possibilities of meaning them, without determinately settling on any one. In Wittgenstein's later method of philosophy, there is no longer any privileged role to be played by a logical symbolism. Rather what we find in its stead are a multitude of examples: examples of the sorts of things that we, when doing philosophy, are tempted to exclaim ('A word stands for a thing', 'Only I can know whether I am in pain' and so on), examples of some of the apparently related sorts of things that we, when not doing philosophy, actually do say or possibly might say ('What does this word stand for?', 'Bring me a slab!', 'I wish I knew what was going on inside his head right now'), and (seeming) examples of situations whose possibility we are invited to imagine (imagine a complete language consisting of only four words, imagine a language which describes your inner experiences and which only you yourself can understand). The point of such interlocking sets of examples and invitations is not to mark out some bright line of grammar which the philosopher is forbidden to cross. Rather, it is to exhibit the grammar of possible ways of speaking to the philosopher, in order to present him with a perspicuous overview of the various possibilities of meaning his words that are genuinely available to him. He is to discover that he can make perfect sense in calling upon the very words he wants to call upon in expressing his philosophical perplexity, although such ways of making sense will not seem to him to answer to what it is he imagined he had originally wanted to say – and to discover that there is nothing further available to mean by his words of a sort which fully answers to what it was he imagined he had originally succeeded in meaning by his words. The aim, both early and late, is to help the reader or interlocutor to see that, until he settles upon one of the determinate things he can mean by his words, he has not yet succeeded in meaning anything by them. Thus, in both Wittgenstein's early and later work, it is left for the reader to discover *for himself* that the problem with his words lies neither in the words themselves, nor in some inherent incompatibility between his words and their context of use, but in his confused relation with respect to his own words.

But then where does the change in his philosophy come? Why do the instruments of logical symbolism become irrelevant to his concern? Why, indeed, does he later look upon such instruments with suspicion? In what way does he take his earlier self to be have been seduced by a vision of the crystalline purity of logic? Why does the manner in which his philosophical work is written undergo such a radical transformation? Why does he come to reject the strategy of offering the reader a single continuous ladder which he is to climb up and then throw away? I believe it is only once one has attained a sufficiently resolute view of where the fundamental continuity in Wittgenstein's philosophy lies that it begins to become possible to make real progress with questions such as these and thus to see where the most profound discontinuities in Wittgenstein's philosophy lie.[18]

Notes

1 Bertrand Russell, 'Introduction' to Ludwig Wittgenstein, *Tractatus Logico-Philosophicus*, trans. C.K. Ogden (London: Routledge, 1922), p. 7.

2 I have attempted such contributions elsewhere. See, for example, my 'The Method of the *Tractatus*', in Erich H. Reck (ed.), *From Frege to Wittgenstein: Perspectives on Early Analytic Philosophy* (Oxford: Oxford University Press, 2001).

3 It should be noted that Gordon Baker is no longer in agreement with the views expressed in the passage I go on to cite here from G.P. Baker and P.M.S. Hacker, *Wittgenstein: Rules, Grammar and Necessity* (Oxford: Blackwell, 1985). Indeed, he tells me that his present reading of Wittgenstein, though differing in detail from the one advanced here, is in broad agreement with the criticisms I here make of 'Baker and Hacker'.

4 Ibid., pp. 39–40, 55.

5 I believe it was Thomas Ricketts who first coined this term and its complement – *resolute readings* – in connection with current debates concerning how to read the *Tractatus*.

6 We might reformulate the distinction here sought, in this case by readers of Wittgenstein's last work, to bring the parallels to standard readings of the early Wittgenstein to the fore, by formulating it as a distinction between that which can be *said within the framework of our linguistic practice* and that which can only be *shown through the employment of framework propositions*.

7 I should perhaps make it clear that, in what follows, I do not mean to impugn the bare claim that Wittgenstein thinks, in *On Certainty*, that there are certain propositions – which he calls 'hinge propositions' – that can be said to be true because they are inferentially related to empirical propositions, although it makes no sense to think of them as either confirmable or disconfirmable (any more than I would want to impugn the claim that he thinks, in the *Tractatus*, that there are certain propositions – a subset of those which he calls 'logical propositions' – that can be said to be true, because they are inferentially related to propositions which make a claim on reality, although it makes no sense to think of them as either confirmable or disconfirmable). What I object to are attempts to extend the category of hinge (or framework) propositions in *On Certainty* to include those forms of words which Wittgenstein seeks to expose as nonsense and to exploit this conflation to give a unitary

account of what sort of thing it is that employments of such forms of words, in common with hinge propositions, are able to 'show'. See Ludwig Wittgenstein, *On Certainty*, ed. G.E.M. Anscombe and G.H. von Wright, trans. D. Paul and G.E.M. Anscombe (Oxford: Blackwell, 1969).

8 Aside from 'The Method of the *Tractatus*', *op. cit.*, n. 2, see also the following papers of mine: 'The Search for Logically Alien Thought: Descartes, Kant, Frege and the *Tractatus*', *Philosophical Topics*, **20** (1), 1991; 'Kierkegaard, Wittgenstein and Nonsense', in T. Cohen, P. Guyer and H. Putnam (eds), *Pursuits of Reason*, (Lubbock: Texas Technical University Press, 1992); 'Throwing Away the Top of the Ladder', *Yale Review*, **79**, (3), 1991; 'Must We Show What We Cannot Say?', in R. Fleming and M. Payne (eds), *The Senses of Stanley Cavell*, Bucknell Review (Lewisburg, PA: Bucknell University Press, 1989); 'Putting Two and Two Together: Kierkegaard, Wittgenstein and the Point of View for Their Work as Authors', in T. Tessin and M. von der Ruhr (eds), *The Grammar of Religious Belief* (London: St Martin's Press, 1995); 'Carnap and Early Wittgenstein on Logical Syntax', in T. McCarthy (ed.), *Wittgenstein in America* (Oxford: Oxford University Press, 2001); 'A Prolegomenon to the Reading of Later Wittgenstein' in C. Mouffe and L. Nagl (eds), *Deconstruction and Pragmatism* (Frankfurt am Main: Peter Lang Press, 2001); 'On Going the Bloody Hard Way in Philosophy', in J. Whittaker (ed.), *The Possibilities of Sense* (New York: Macmillan, 2003); 'What "Ethics" in the *Tractatus* is *Not*' in D.Z. Phillips (ed.), *Wittgenstein on Ethics and Religion* (Aldershot: Ashgate, 2005).

9 'The Method of the *Tractatus*', *op. cit.*, n. 2.

10 In the entry entitled 'Nonsense' in Hans-Johann Glock, *A Wittgenstein Dictionary* (Oxford: Blackwell, 1996), pp. 259–60), Glock is helpfully explicit in attributing the substantial conception of nonsense to the *Tractatus*.

11 In claiming that the *Tractatus* is to be seen as resolving a tension in Frege's thought (between these two different conceptions of nonsense), I touch on an interpretative question about how Frege is to be read. I mean to take sides on this question only in so far as it bears on the claim that Wittgenstein can be fruitfully read as having read Frege in certain ways.

12 'The book will, therefore, draw a limit ... not to thinking, but to the expression of thoughts; for, in order to draw a limit to thinking we should have to be able to think both sides of this limit (we should therefore have to be able to think what cannot be thought).' (Ludwig Wittgenstein, *Tractatus Logico-Philosophicus*, trans. C.K. Ogden (London: Routledge, 1922), Preface, p. 27.)

13 'The limit can, therefore, only be drawn in language and what lies on the other side of the limit will be *simply* nonsense' (Preface, my emphasis). As noted above, the 'simply' here goes with the 'only' over which we paused earlier: 'Every possible proposition is legitimately constructed, and if it has no sense this can *only* be because we have given no meaning to some of its constituent parts' (TLP 5.4733, my emphasis).

14 Commentators fail to notice that what Wittgenstein says in TLP 6.54 is not: '*all* of my sentences are nonsensical' (thus giving rise to the self-defeating problematic that Geach has nicely dubbed *Ludwig's Self-mate*). Rather, TLP 6.54 characterizes the way in which *those* of his propositions which serve as elucidations elucidate. He says: 'my sentences serve as elucidations in the following way: he who understands me recognizes them as nonsensical.' The aim of the passage is *not* to propose a single all-encompassing category into which the diverse sorts of proposition which comprise the work are all to be shoehorned, but rather to explicate how those passages of the work which succeed in bearing its elucidatory burden are meant to work their medicine on the reader.

15 Ludwig Wittgenstein, 'Lectures on Personal Experience', ed. C. Diamond, unpublished, 1935.

16 See Stanley Cavell, *The Claim of Reason: Wittgenstein, Skepticism, Morality and Tragedy*, (Oxford: Oxford University Press, 1979), pp. 344ff.
17 See my 'Wittgenstein on Meaning and Use', *Philosophical Investigations*, **21** (3), 1998, pp. 222–50.
18 For further discussion of this topic, see J. Conant and C. Diamond, 'On Reading the *Tractatus* Resolutely' (in M. Koelbel and B. Weiss, *The Lasting Significance of Wittgenstein's Philosophy*, Routledge: London, 2004).

– A prolegomenon to the possibility of a proper understanding... (167)

Chapter 9

Transcendence and Contradiction in the *Tractatus*

Barry Stocker

Propositions*

> The transcendental dialectic will therefore content itself with uncovering the illusion in transcendental judgements, while at the same time protecting us from being deceived by it; but it can never bring it about that transcendental illusion (like logical illusion) should even disappear and cease to be an illusion. For what we have to do with here is a **natural** and unavoidable **illusion** which itself rests on subjective principles and passes them off as objective, whereas logical dialectic in its dissolution of fallacious inferences has only to do with an error in following principles or with an artificial illusion that imitates them. Hence there is a natural and unavoidable dialectic of pure reason, not one in which a bungler might be entangled through lack of acquaintance, or one that some sophist has artfully invented in order to confuse rational people, but one that irremediably attaches to human reason, so that even after we have exposed the mirage it will still not cease to lead our reason with false hopes, continually propelling it into momentary aberrations that always need to be removed.[1]

The *Tractatus* can, and should, be conceived as a continuation of the Kantian project of exposing and dissolving dialectical illusion, on the assumption that dialectic is a natural product of the theoretical interests of pure reason and that it is unavoidable.[2] Wittgenstein takes up the Kantian themes and takes them further. The *Tractatus* is a work which is undecidable with regard to whether it is an example of the dialectic of clarification, or the putting forward of theses which uniquely resist dialectic. A series of numbered propositions claim to give an apodictic account of the limits of philosophy and the logical space of the world. Careful reading strongly suggests extreme equivocation and even contradiction in the theses, leading up to the final claim that the propositions are nonsense to be overcome. The *Tractatus* puts forward a seemingly unchallengeable set of propositions in order to show their limits and overcome them.

In the *Tractatus* Wittgenstein is concerned with the correct form of propositions in the critique of language, a critique restoring language to a logical purity that is transcendental, and as such is a mirror of the world.[3] As logical purity is inherent in propositions, the *Tractatus* necessarily declares itself to be superfluous and engages in philosophy at its limits where contradiction arises. Contradiction in Wittgenstein should be seen as more than a simple assertion in TLP 6.54, that his propositions, when understood, will be understood as nonsensical. The meaning of 'nonsensical' has to be determined in Wittgenstein, and this can best be done with reference to the questions of transcendence and contradiction in philosophy, as raised by Kant.

The *Tractatus* nears its end with the claim that: 'There are, indeed, things that cannot be put into words. They *make themselves manifest*. They are what is mystical. [*Es gibt allerdings Unaussprechliches. Dies z e i g t sich, es ist das Mystische*]' (TLP 6.522).[4] Wittgenstein builds up to the claim that there is a limit to propositions, which is the subject matter of the *Tractatus*, but which cannot be said in the work. Earlier parts of the book refer to the forms of propositions, language and the world. The idea of saying and showing in the proposition is introduced: 'A proposition *shows* its sense. A proposition *shows* how things stand *if* is true. And it *says that* they do so stand. [*Der Satz z e i g t seinen Sinn. Der Satz z e i g t, wie es sich verhält, wenn er wahr ist. Und er s a g t, d a s s es sich so verhält*]' (TLP 4.022).[5] There are two aspects to the proposition, language and logical form: saying and showing. Showing is a showing of the form of a fact; saying is the assertion that the fact is what is the case and not just a possibility. The diremption of the proposition between saying and showing must have strong and strange effects on what a proposition is, and what language or logic might be.

The philosophical results of this diremption are explored as in:

> Propositions cannot represent logical form: it is mirrored in them. What finds its reflection in language, language cannot represent. What expresses *itself* in language, *we* cannot express by means of language. Propositions *show* the logical form of reality. They display it. [*Der Satz kann die logische Form nicht darstellen, sie spiegelt sich in ihm. Was sich in der Sprache spiegelt, kann sie nicht darstellen. Was s i c h in der Sprache ausdrückt, können w i r nicht durch sie ausdrücken. Der Satz z i e g t die logische Form der Wirklichkeit. Er weist sie auf*]. (TLP 4.121)

The translators in both the Pears and McGuinness and Ogden translations are determined to say that plural propositions show the form of logic, although Wittgenstein suggests that the single proposition displays logical form. This is not exactly bad translation, but it overlooks the possibility that Wittgenstein wished to emphasize that the single proposition contains logical form, so that the possibility of there being a proposition at all is the significant aspect, and we do not need to examine the empirical conditions of individual propositions in order to discern the logical form of the proposition.[6] In other words, we should not look for propositions to exemplify logical form and give examples of it; the possibility of a proposition at all is the possibility of logical form.

There is a paradoxical form of expression here, typical of the *Tractatus*, which has not been sufficiently emphasized in previous commentary. Logical form is expressed 'in' (*in*) language but cannot be expressed 'by', or 'through' (*durch*), language. The difference between 'in' and 'by' is accompanied by the difficulty of the difference between 'reflection/mirroring' and 'representation', or between 'showing/displaying' and 'by'. Wittgenstein assumes the possibility of a showing or display which expresses in language, whereas expression by language is impossible. The form which is expressed can be reflected/mirrored but not represented. The difference appears to be that between the saying of a proposition and what is shown in the prepositional structure of a proposition, in a proposition being what a proposition is. This means that language can show, but not say, what its own form is, so that the capacity of language to say rapidly and naturally leads to illusion and error. The role of saying is assertion, but assertion keeps bursting beyond its proper role into propositions about form. Even to say that propositions of this kind are illegitimate is an illegitimate use of propositions: philosophy must be dialectical, as Kant suggested it must be in part, but for Wittgenstein dialectic overwhelms philosophy which, as philosophy, must be dialectic not in part, but in the whole of what has been previously recognized as philosophical writing.

Saying + showing

The status of reflection is reinforced in a later proposition: 'Logic is not a body of doctrine, but a mirror-image of the world. Logic is transcendental. [*Der Logik ist keiner Lehre, sondern ein Spiegelbild der Welt. Der Logik ist transzendental*]' (TLP 6.13). Propositions mirror logical form rather than representing it. Logical form presumably cannot be (re)presented. Logical form itself mirrors the world. Mirroring appears to be identical with showing (as opposed to saying) and is the limit of saying. However, if we go back to TLP 3.031:

> It used to be said that God could create anything except what would be contrary to the laws of logic. – The truth is that we cannot *say* what an illogical world would look like [*Wir könnten nämlich von einer „unlogischen" Welt nicht s a g e n, wie sie aussähe*].[7]

If we cannot say what an illogical world is like, we also cannot say what a logical world is like. This leans towards suggesting that propositions/language/logical form is saying, and that showing is a sphere within saying. In this case, language might be taken as assertion in which assertion does not just give the affirmative or negative to a possible fact, but also gives, shows, the possible facts. Language itself is caught in an unstable diremption between saying and showing as, at differing points in the *Tractatus*, language is taken over by either saying or showing.

TLP 4.022 suggests that there is nothing outside what can be shown. Saying is alluded to as where we reach beyond showing or as simple affirmation or negation of what is shown – that is, whether the proposition refers to an actual state of affairs and not just a possible state of affairs. Since saying takes over language in TLP 3.031, there is an implication that all language contains metaphysical propositions, although Wittgenstein suggests that there could be a language of logical propositions (TLP 6.122) which would be clear in themselves. If we consider saying as just affirmation or negation, then this fits into the pure language of logical propositions; if saying is the transcending of what is shown, then we are in the territory of the dialectic of philosophical language. The *Tractatus* operates between the poles of pure logic and the attempt to speak the unsayable.[8]

Pure logical propositions would dispense with the need for propositions of logic (TLP 6.121), which say nothing. If logic is tautological, as Wittgenstein suggests in the *Tractatus*, then the status of language becomes very strange. Language is also said to be logically correct, but if it follows logical form it is the statement of tautologies; anything else must be showing. The capacity of language to say is restricted to affirmation or negation, and everything else is pure showing.

The very attempt to say anything about logical laws, or general logical objects, is metaphysical, on the other side from a metaphysically pure language; but it is not clear that there is a sustainable Tractarian distinction between metaphysics and normal propositions. The saying in logical propositions is a degree zero at the limits of philosophy where a proposition says in showing, so that there is recognition of logic but no recognition of laws of logic.

The proposition does not show states of affairs in some direct way. Wittgenstein is not satisfied with the idea of sentences corresponding directly with arrangements of objects. They do so correspond, but only on the condition of representing possible states of affairs in logical space, in a logically structured proposition. What the proposition represents is the possibility of a state of affairs as part of the transcendental and logical structure of the world, the sentence does not directly represent a state of affairs.

What can be represented is sense: 'What a picture represents is its sense' (TLP 2.221). A proposition includes all that the projection includes, but not what is projected: 'A proposition, therefore, does not actually contain its sense, but does contain the possibility of expressing [*ihn auszudrucken*]' (TLP 3.13). Sense is what the proposition can refer to, but only through the possibility of expressing sense. The world has a logical structure with relations between objects: propositions are what show sense – the logically possible relations.

Propositions mirror logical form: logical form mirrors the world. Logical form is what is representable. What is contained in the logical form of the proposition is the possibility of expressing sense. Sense, in this case, is what is expressed through the possibility of a propositional form which mirrors transcendental logical form, which in turn mirrors the world. Sense is determined by the world in the possibility of its expression. There is an inner transcendental logical form, which in turn mirrors the world. Sense is determined by the world in the possibility of its expression. There is an inner transcendental (or pre-transcendental?) determination of sense and the world. The picture theory of meaning specifies this transcendental determination:

> The fact that the elements of a picture are related to one another in a determinate way represents that things are related to one another in the same way. Let us call this connexion of elements the structure of the picture, and let us call the possibility of this structure the pictorial form of the picture. [*Dieser Zusammenhang der Elemente des Bildes heisse seine Struktur und ihre Möglichkeit seine Form der Abbildung*]. (TLP 2.15)[9]

The possibility of sentences referring to objects is questioned. It is not the reference to objects which determines language, it is the logical possibilities inherent in it which are a question of the relations within a proposition. The logical space of the world is the same as the logical space of the propositions; therefore propositions can have a connection with the states of affairs in which objects are arranged. This allows the proposition to be a picture, though a picture of relations not objects. The bearer of meaning in language is a proposition not words, as Frege[10] had already suggested, and Russell had emphasized.[11] Going beyond Frege, Wittgenstein denies that propositions can be complex names;[12] or that, as Russell suggest, there can be logically proper names referring propositions to indescribable objects of perception.[13] Wittgenstein emphasizes how this leads us to define meaning in terms of the giving of the world through logical structure. An ambiguity appears between propositions as pure logical possibility and as what refers to actuality. The use of affirmation and negation does not solve this problem, as the possibilities of affirmation are in logical form; they do not pre-exist language in the world. The effect of the *Tractatus* is to create a transcendental world of language form, which could mirror the world and could be what constitutes the world for us. Logical form, inherent in language, reflects the world and propositions reflect logical form. Therefore the primary aspect of a proposition is that it reflects what reflects the world. The proposition shows states of affairs in so far as it reflects the logical possibility of a proposition. It can show a state of affairs because of the underlying common logic of states and affairs and propositions, which leaves questions of logical form prior to the possibility of displaying states of affairs. Logical form transcends, states of affairs are displayed on the condition that there is what transcends.[14]

Objects

The beginning of the *Tractatus* itself seems to be constituted by ambiguity. We should compare TLP 1.1, 2 and 2.01:

> The world is the totality of facts, not of things. (TLP 1.1)
>
> What is the case – a fact – is the existence of states of affairs. (TLP 2)
>
> A state of affairs (a state of things) is a combination of objects (things). (TLP 2.01)

It seems that Wittgenstein wants to say both: the world is analyzed into facts which excludes the possibility that it can be analyzed into things; and that facts can be analyzed into things, in which case it must be the world that can be analyzed into things. This contradiction seems to have been largely overlooked in the commentary and the possibility of the *Tractatus* beginning with a contradiction has not been acknowledged. But it must be the case here that either Wittgenstein has indulged in clumsy unconsidered vagueness, or that the *Tractatus* is a work which begins in paradox and deals with philosophical paradox in its means of exposition.

It cannot be the case that the world is made up of facts instead of things, but that facts are combinations of things in states of affairs. It could be said that the world is not directly made up of things, but is rather made up of facts composed of things, but this is in opposition to the absolute nature of the opening proposition and would still leave a strong impression of paradox in any case.[15] Later propositions, which state that it is nonsense to talk about 'logical objects in general', confirm that paradox and contradiction must be in the *Tractatus* as its mode of exposition.

According to propositions 2.021 to 2.031, objects are both form and substance, which combine form and content. Form is colour, time and space. Objects are colourless, so objects are formless. Objects are subsistent, unalterable and absolutely independent in the *Tractatus*, while being related in logical possibility to other objects, through internal logical possibilities. Objects, as what necessarily contain logical possibility, become the form uniting all possible worlds. This is logical form, also referred to as transcendence. Here, objects are the logical possibility of objects, not particular objects, and are not explicitly defined in any case. Propositions express relations between objects, which are essentially expressions of internal relations. However, it is not possible to *say* what are the logical forms governing propositional form. This surely leaves colour, space and time as transcendental forms of objects. They are forms external to objects themselves and are therefore presumably rooted in how we perceive objects, and must precede logic. Logic as transcendental form seems to be replaced by time, space and colour. As objects do not have any existence outside the transcendental forms of logic posited in language, according to the *Tractatus*, they reproduce the Kantian account of transcendental forms and ideas. The substance determined by objects is itself both composed of objects and what is external to objects. The ambiguity may be further revealed in comments on form and structure: 'Form is the possibility of structure. [*Die Form ist die Möglichkeit der Struktur*]' (TLP 2.033); 'The structure of a fact consists of the structures of states of affairs. [*Die Struktur der Tatsache besteht aus den Strukturen der Sachverhalte*]' (TLP 2.034). The notion of form, linked sometimes with what is external to the object, now appears to be the possibility of structure itself, which allows the creation of facts composed of states of affairs. There is a slippage between what belongs to objects and what belongs to logical form. The world supposedly consists of facts, not things. Facts are described in terms of structure and form, but clearly these belong to the internal properties of objects. Facts are what express the internal relations of objects. Wittgenstein slips between opposing facts to things, and implying that facts are combinations of things. Things are what contain logical possibilities and determine substance, in any possible world, which looks like undermining the distinction between facts and things. A thing cannot be conceived outside its factuality, its internal properties regulating relations.

There is a reworking here of Kantian transcendental arguments about things-in-themselves and the absolute forms of perception or judgement. In Kant, space and time are absolute forms of perception, and judgements of the objects of experience necessarily refer to a transcendental self, concepts of experience and ultimately ideas of reason. For Kant, ideas of reason tend to create paradoxes and contradictions as they reach beyond the objects of cognition. Given the necessary existence of a faculty of reason, detached from empirical objects, this seems to be a necessary source of illusion in philosophy. Following this, it might be considered that philosophy is intrinsically in the realms of the paradoxes of reason. Unlike Kant, Wittgenstein implicitly suggests that any writing about objects and transcendental conditions must indeed be governed by paradox and contradiction.

The Kantian structure can be further demonstrated with some clear references to Kant in the Tractarian propositions. TLP 4.0412, for example, follows comments about the necessity of mathematical multiplicity in formal symbolism, which cannot be the subject of depiction (*abbilden*) – presumably a form of showing (*zeigen*) – but enters into depiction:

> For the same reason the idealist's appeal to 'spatial spectacles' is inadequate to explain the seeing of spatial relations, because it cannot explain the multiplicity of these relations. (TLP 4.0412)

This remark contains a strong suggestion of affirmation by denial. Why mention Kantian absolute forms of perception, unless they tell us something of Wittgenstein's underlying philosophical claims? The implication of this proposition is that there are idealist spectacles, but they are multiple. Logic itself is the expression of absolute forms of perception through the multiplicity of relations. Wittgenstein seems to be claiming to be non-idealist because his absolute forms of perception are multiple, but this does not justify the said claim.

Wittgenstein elsewhere quite explicitly invites us to consider his notion of logical relations as an extension of Kantian forms of perception:

> Just as we are quite unable to imagine [*denken*] spatial objects outside space or temporal objects outside time, so too there is *no* object that we can imagine [*denken*] excluded from the possibility of combining with others. (TLP 2.0121)

What should we now make of TLP 6.3611 which denies the absolute form of time and belongs with many propositions denying absolute forms of perception and laws of causality. TLP 6.36111 denies the force of the Kantian example from the *Prolegomena*[16] of the impossibility of replacing the left hand with the space of the right hand, although they are congruent. Kant used this to argue that that there must be absolute space, since nothing distinguishes the two hands except their position in space. This is the only explicit reference to Kant in the *Tractatus*, and seems to distance Wittgenstein from Kant, but this is misleading. Wittgenstein claims that Kant's problem can be solved by referring to a fourth dimension and by the real congruity of objects, even if they cannot be made to coincide within a limited number of dimensions.

Again, this is a denial which affirms a relation with Kant. There is no undermining of Kant in the existence of the fourth dimension; the problem could still be repeated in a dimension. In any case, Wittgenstein's claim that the gloves will be congruent in the fourth dimension misses the point. Kant states that they are congruent in three dimensions, but their non-intersubstitutability shows that space is a quality in itself. Wittgenstein also claims that substitutability is irrelevant compared with congruence, but this is not explained. Kant is rejected at the very moment that Wittgenstein is assuming a Kantian framework of transcendental forms. Wittgenstein does not so much replace Kantian idealism with post-Fregean philosophy, as expose the Kantian roots of Frege, the necessarily Kantian roots of any discussion of judgement and absolute forms. There is no consistency in Wittgenstein's repudiation of absolute forms of perception. A rejection of 'idealism' is asserted at the same time as an argument is put forward on the border of idealism and logical form, which suggests that the discussion of logical form leads both to idealism and nonsense. It is not just the Kantian discussion of the forms of objects and the world which leads to nonsense. In his own words, so does Wittgenstein's own proposition on the forms of time, space and colour, and what objects are in general. Is not Wittgenstein's condemnation of Kant's spectacles itself a condemnation of Wittgenstein's own glasses?

Kantian overtones are clear through the *Tractatus*, as in TLP 2.1: 'We picture facts to ourselves. [*Wir machen uns Bilder der Tatsachen*]'. Facts are what we give to ourselves. This is followed by the elaboration of the 'picture theory of meaning' which connects with Wittgenstein's discussion of thought, propositions and logic. 'Pictorial form is the possibility that things are related to one another in the same way as the elements of the picture' (TLP 2.151). Pictures are what we make for ourselves and the possibility of the form of pictures is the possibility of the relations between things. Things and their relations are implicitly understood as what is known through the faculty of logic/propositions/pictures/thoughts; they do not have a separable existence from what comes within the transcendental limits of our picturing, representation and thought. For some commentators, Wittgenstein succeeds in overcoming the idealist problem through identifying the world with language.[17] However, even if Wittgenstein's remarks on language might lead us in such a direction, this does not mean that the problem is resolved. The problem must be repeated, which is particularly clear when we remember that Wittgenstein refers to propositions as representations. What is most usefully learned from the *Tractatus* is not that metaphysical problems are resolved in language, but that metaphysical problems have necessary origins *in* language.

Dialectic

The discussion of objects in the *Tractatus* can be considered in the light of Kant's comment from the beginning of the 'Transcendental Dialectic' in the *1st Critique*:

> However it may be with the possibility of concepts from pure reason, they are not merely reflected concepts but inferred concepts. Concepts of the understanding are also thought *a priori* before experience and on behalf of it; but they contain nothing beyond the unity of reflection on appearances, insofar as these appearances are supposed to belong necessarily to a possible empirical consciousness. Through them alone is cognition, and determination of an object, possible. They also first give material for inferring, and no *a priori* concepts of objects precede them, from which they could be inferred. On the

> contrary, their objective reality is founded solely on the fact that because they constitute the intellectual form of all experience, it must always be possible to show their application in experience.[18]

For Kant, there can be no *a priori* concept of the object which precedes the concepts of understanding applicable to experience. When considering the concepts of pure reason, we must exclude concepts which can only be legitimately considered as belonging to understanding and applicable to experience. This brings Kant into ambiguous territory, the territory of dialectic. According to Kant, pure reason cannot be grounded except as what is applicable in experience. The point of a critique of pure reason is for pure reason to limit itself to experience by excluding its innate tendencies towards metaphysical dogmatism. The situation is that there is something before experience, theory or pure reason concerned with what is outside, or before, the objects of time and space, but that there can be no concepts or judgements which are legitimately independent of experience in time and space. The creation of systems, which are not properly founded on experience as constituted by the concepts applicable to time and space, is the creation of metaphysical illusion. The purpose of the *1st Critique* is to show us that illusions are illusory and return us to concepts usable in experience. The whole 'Transcendental Dialectic' of the *1st Critique* is devoted to showing the type of errors which arise from taking concepts beyond their proper limits in 'rational' psychology, cosmology and theology.

Dialectic is error to be overcome according to Kant but, on the other hand, 'dialectic' is used in a way which suggests that it is the positive activity of showing the limits to concepts and their real relations. This ambiguity is the ambiguity of philosophical writing in Kant. How can there be a writing of 'critique' as doctrine or dialectic, when the use of concepts outside their application is intrinsically illusory? What is the difference between illusory dialectical philosophy and the philosophy which establishes doctrine? How can doctrine avoid the illusions of pure concepts taken as independent of experience? From Kant's position does it not become inevitably the case that all philosophical writing is dialectic, the overcoming of illusion? Is Kant not most dialectical in the bad sense in his doctrine? Kant's own distinction between critique and metaphysics implicitly questions the possibility of metaphysical system, and certainly renders it secondary to the purity of judgement in pre-systematic reason, which itself must be defined with regard to dialectic.

All these ambiguities and questions are present in the *Tractatus*, which works through the paradox of philosophical writing in relation to the priority of system and concepts outside experience. Wittgenstein's discussion of objects both establishes a 'transcendental' object of logical space, with properties of time, space and colour and condemns the idea of an *a priori* concept of object, apart from what appears in logical notation and language as expression of logic. The unavoidable dialectic in the *Tractatus* appears in the distinction between saying and showing. The distinction is not that between factual saying and mystical showing, it is a distinction within the proposition between the assertion of truth or falsity and the representation of a possible state of affairs. The truth or falsity of a proposition is said; the possible state of affairs is shown. Saying has a difficult status in the *Tractatus* since there is very little which can be said. It is reduced to the representation of possible states of affairs, tied to states of affairs through a common structure which enables the proposition to picture the state of affairs, and which is logic.

Logical form itself cannot be said, nor can the relationship between proposition and states of affairs be said. The bases for the solution of mathematical problems are not said; they are contained in the *intuitions* of language (TLP 6.233), strongly suggesting that scientific and mathematical law, in general, is not said but shown from the intuitions contained in language – logical form. Language itself contains logical form and is where there is logic. We do not need to say laws of logic, as logic is present in the correct ordering of symbols and their connections. A specific logical notation makes clear the logical nature of language. Language does not need to be added to or revised, in this perspective:

> In fact all the propositions of our everyday language, just as they stand, are in perfect logical order. (TLP 5.5563)

> It follows from this that we can actually do without logical propositions; for in a suitable notation we can in fact recognize the formal properties of propositions by mere inspection of the propositions themselves. (TLP 6.122)

These remarks leave an area of ambiguity. Is it the case that language in itself is logically in order, or do we need a correct symbolism in a rearranged language in order to show logical form? If language is logically in order on its own, why do we need a formal notation at all?

However, there is a philosophical activity of clarifying propositions and there are mistakes in language, which make it necessary for there to be a text such as the *Tractatus*. The *Tractatus* claims to reform language only in restoring it and showing the commentary of saying what language is to be superfluous:

> The correct method in philosophy would really be the following: to say nothing except what can be said, i.e. propositions of natural science – i.e. something that has nothing to do with philosophy – and then, whenever someone else wanted to say something metaphysical, to demonstrate to him that he had failed to give a meaning to certain signs in his propositions. (TLP 6.53)

In this case, it must be wondered why language can lead to illusion, since in itself it is perfectly logical. The *Tractatus* could be a showing of how language says the illusory, because it contains necessary possibilities of ambiguity, just as it contains the logical space of objects and propositions. Logical space is the limit of what is, which is what *we* represent (this point, the *we*, is clear on a number of occasions), and is the logic implicit in facts, propositions and language. In which case there is still the problem of how there can be illusion in language.

Logical space contains the logical possibility of failures of language to mean: *bedeutungslos*, *unsinnige*, and *sinnlos* propositions. These three failings of meaning or sense are important in the *Tractatus*, particularly with regard to the distinction between *unsinnige/sinnlos* and *bedeutungslos*, which look identical in Wittgenstein's discussion, although he does not directly suggest such a thing. Nonsense is famously Wittgenstein's own epithet with regard to the *Tractatus*. *Bedeuteungslos* refers to the failure to use a concept which can be applied – its appearance in a context where it cannot be used. At this point, the sign is failing to relate to symbols (TLP 4.461). These will now be referred to together as non-meaningful (*bedeuteungslos*). Examples include contradictions and tautologies; and there is the suggestion that Occam's razor is a rule to exclude such cases (TLP 3.328, 5.47321) (*bedeuteungslos*). Nonsense (*Sinnlos*) is something more important and more ambiguous. It can be linked to the non-meaningful – nonsense can appear from a non-meaningful hypothesis taken as a premise. However, nonsense suggests something more systematic than the non-meaningful. It is the use of misleading concepts with no basis and which create illusion – for example, the general concept of an object (TLP 4.1272) (*unsinnige*); the statement that two objects are identical (TLP 5.5303) (*unsinn*); the addition of a sign for a proposition to a proposition (TLP 5.5351) (*unsinn*). Nonsense refers to errors in the use of language and logic, in the representation of objects, facts, substance, and structure of a kind which leads to misleading statements on these topics. But all statements on these topics are nonsense; to say anything about them is to talk nonsense. This might be considered in the light of Wittgenstein's comment in the preface:

> Thus the aim of the book is to draw a limit to thought, or rather – not to thought, but to the expression of thoughts: for in order to be able to draw a line to thought, we should have to find both sides of the limit thinkable (i.e. we should have to be able to think what cannot be thought). It will therefore only be in language that the limit can be drawn, and what lies on the other side of limit will simply be nonsense.[19]

The preface declares the goal of excluding nonsense from philosophy; TLP 6.54 announces that the *Tractatus* is nonsense. It follows that the *Tractatus* attempts to draw a limit to thought, not just to the expression of thought, and that it commits the fundamental error of attempting to think both sides of the limits of language. It is stated in the *Tractatus* that philosophy is the activity of clarifying propositions, with the admission that this would not be widely recognized as a philosophical activity. What the *Tractatus* has perhaps done is discover that the clarification of propositions cannot avoid falling into statements about the conditions of propositions. Such statements must draw a limit to propositions and therefore must be nonsense.

The nonsense is apparent in propositions 1 to 2.01, or should be apparent if the reader concentrates on the paradoxical nature of the *Tractatus*. The paradox is that philosophy should be only showing but is full of saying, and that to say this is to go beyond the bounds of what can be said. It is important to concentrate on the paradoxes of the *Tractatus* and on what is immanent in language. Language contains the *a priori* possibility of logic and of scientific law. Laws of logic are superfluous additions and causal laws of nature are superstition. Logical space is contained in language, which is why philosophy is the clarification of propositions. What this leaves aside is the discussion of the status of language. Language stands in for Kantian transcendental conditions of knowledge; it is where there is *a priori* judgement in the structure which constitutes language. Wittgenstein's *Philosophical Investigations*[20] take up this task and therefore, as TLP 5.5563 strongly suggests, constitute a continuation of the *Tractatus* – a continuation in which Wittgenstein tries to overcome the contradiction of saying the unsayable, by discussing the conditions of saying.

Wittgenstein's philosophical interests and goals are not stated exactly. Therefore it is difficult to state with certainty that the *Tractatus* is a text which shows itself to be nonsense, with no theses to defend. This has led some to stoutly defend the notion that the *Tractatus* is simply a set of meaningful propositions about propositions, and to claim that the 'therapeutic' approach to the *Tractatus*, exemplified by Diamond[21] and Conant,[22] is in danger of simple rejection – an absolute denial of the theses contained therein. It is the systematic reading of the *Tractatus* (Stenius and Hacker[23]) that is in greater danger of ignoring the way in which the theses are not only suggested to be nonsense in TLP 6.54, but also how the text from the preface on revolves round the paradoxes and contradictions of the project suggested. The distinction between the two points of view is not absolute. Diamond's 'Throwing Away the Ladder'[24] is mostly concerned with the theory of sentences in Wittgenstein, compared with Frege and Russell, unlike Conant, although Conant does have a great deal to say about the Fregean influence.[25] In this argument, Diamond suggests that the *Tractatus* should be read with regard to two themes. One is that in contrast to Hacker's reading,[26] there is no reference to ontological reality outside sentences, only what the form of sentences tells us about reality, following the precedent of Frege. Another is that general statements about logical form are nonsensical because logical form is what shows itself. This still leaves Diamond's reading demonstrating a theory of sentences, truth functions and so on. There is also a failure to engage with the dialectical nature of Wittgenstein's philosophical writing, but, given that Diamond is mostly concerned with Wittgenstein's roots in Frege and Russell, this is not surprising.[27] Conant gives more consideration to Kierkegaard and dialectical style (although he does not use the phrase) in various places including 'Throwing Away the Top of the Ladder'. There is a shared objective, with Diamond, of emphasizing that we must take seriously the *Tractatus*'s self-definition as nonsense, and both authors have helped to reorientate *Tractatus* studies, but from different directions.

Reid's 'Wittgenstein's Ladder'[28] criticizes Conant and Diamond as if they formed one philosopher, for not seeing that nonsense in the *Tractatus* is given an absolute definition abandoned in the *Philosophical Investigations*, and that the absolute definition of nonsense leads to contradiction in asserting propositions which are said to be nonsense. Reid's dismissal indicates her own failure to appreciate the possibility of writing at the limits of sense and contradiction, and why this strategy might be adopted. Conant is also accused of emptying the

Tractatus of content in Lippitt and Hutto's 'Making Sense of Nonsense: Kierkegaard and Wittgenstein'.[29] This emphasizes the idea that theses, showing the limits of what can be thought, are advanced in the *Tractatus* which are not wholly rejected, and an account of Kierkegaard is offered to back this up. Their reading shows a 'dialectical' approach, but still tends to assume that the propositions of the *Tractatus* can accepted as such, without being affected by the 'framing' comments on the limits of thought and what can be said. It must be emphasized that the *Tractatus* puts all its propositions under question without dismissing them as 'absolute nonsense', which is in fact what the Lippitt and Hutto reading of Kierkegaard would lead us to expect.

Law

The necessity Wittgenstein eventually felt to move towards the *Investigations* arises from a language in the *Tractatus*, which contains *a priori* conditions in the most dogmatic manner. The dogmatism itself just leads into the paradoxes of seeing language as a place of an *a priori* form of logic and scientific law while condemning general laws as superstitious and useless, leading to nonsensical claims about objects as a universal concept. This can only be managed from assertions about intuition in language and its necessary form, which are not justified or explained beyond universal assertion. The universal assertion is itself the condition of a saying which says too much. 'Showing' in Wittgenstein suggests a way out of the paradoxes and nonsense of universal philosophical propositions, but leaves us still having to say what philosophy is and what illusion is. In what way can we assume that saying and showing are clearly distinct and opposite in any case?

Saying shows its limitations and paradoxes, returning us to pure showing. But what is this showing? It must contain the same contradiction as Wittgenstein's language, and also contradictions which arise from the strategy of 'nonsense'. Showing must contain the possibility of saying what can only be shown and of the illusions of language (which is supposed to be alright on its own) about what should only be shown. The relation between showing and language is unclear: if language shows in properly formed propositions, what is the distinction between saying and showing? The separation between saying as assertion and showing as representation relies on a bizarrely reduced notion of both saying and showing in an impossible opposition, which is ultimately contentless and non-meaningful in a way that both fits in with the strategy of nonsense and questions it. Language has become essentially showing rather than saying; saying is the statement of a truth value. One advantage of this for Wittgenstein is that it releases us from the Fregean error of turning the truth value into the reference for a proposition,[30] and it is this kind of logical object that Wittgenstein is constantly trying to avoid and condemns most strongly as illusory. This condemnation of the logical object extends to the condemnation of any statement about logical objects in general as illegitimate saying. The artificiality of the Fregean device and its possible metaphysical commitments is extended into any discussion of meaning, reference, laws and objects of logic. We can give the general form of the proposition but cannot give its general properties (TLP 4.5). This all leads us to the suspicion that, for Wittgenstein, all saying is nonsense based on the non-meaningful use of concepts.

The correct use of concepts excludes the possibility of going beyond logically correct statements, although some of Wittgenstein's claims might lead us to think that it is not possible to speak illogically. Showing is what cannot be illogical because it is depiction according to logical form, but it *is* a concept of language at a degree zero where it is vanishing. The opposite pole of the zero-method is highlighted in TLP 6.121:

> The propositions of logic demonstrate the logical properties of propositions by combining them so as to form propositions that say nothing. This method could be called a zero-method. In a logical proposition, propositions are brought into equilibrium with one another, and the state of equilibrium then indicates what the logical constitution of these propositions must be.

Showing is the impossible moment in which language has achieved total transparency in sharing the logical structure of world, formed according to substance and logical space. Showing and saying are two poles of language: its absorption into what it depicts; a statement about the universe. Language as saying something about something threatens to fly free of universal substance.

There is a sphere outside the world, transcendental values of ethics and aesthetics,[31] but these can still only be shown. It is in the transcendental that the purity of showing becomes more insistent. The transcendental values are given a similar place to the logical structure, substance and essence of the universe. The opposition between transcendental values and the logical statements of world structure is absorbed in their common position. In this case, the subject matter of the *Tractatus* should occupy the same place and can only be what is shown and not said, which is precisely what TLP 6.54 and the preface suggest are absolute transcendental statements. So, for Wittgenstein, there is the space of transcendental statements which are necessary and contradictory. The Kantian claim that reason is intrinsically governed by the desire to go beyond its proper limits and make contradictory statements is taken further.[32] There are no transcendental statements of pure reason which are not contradictory according to Wittgenstein: language is intrinsically riven by the opposition of saying and showing. That is not an explicit statement of Wittgenstein, but it is an unavoidable result of the *Tractatus*.

Other aspects of the *Tractatus* lead us away from the purity of transcendental critique into territory which looks suspiciously like the return of metaphysical hierarchies. The account of logical form (nonsensical though it must be) is used as the basis of scientific law in TLP 6.32–6.341. The discussion of laws of nature leaves ambiguity, particularly in TLP 6.3211, which suggests that law of least action must be logical because it is known *a priori*. It is not explained why the law of least action is logical, except on the basis that there is *a priori* knowledge of a logical form (6.33). There is strong ambiguity here. Does Wittgenstein wish to suggest that the most fundamental laws of physics are *a priori* (as Kant does[33])?

This would be in manifest contradiction with the proposition preceding the discussion of law: 'The exploration of logic means the exploration of *everything that is subject to law*. And outside logic everything is accidental' (TLP 6.3). This strongly suggests that what is not logical is accidental and not law-governed. However, it leaves a second possible meaning – namely, what is law-bound in nature is bound by logic in its lawfulness. Two extreme positions appear. One position is that propositions can be divided absolutely between propositions of logic and propositions of fact. The second position is that fundamental scientific laws have something *a priori* and logical about them. The second position is compatible with the first if the logical, *a priori* aspect of law is interpreted as compatibility with logical form (as TLP 6.33 and 6.34 seem to suggest). However, what Wittgenstein says in TLP 6.3211 strongly suggests that there is a logical necessity in the law of conservation, beyond mere formal compatibility with logical space – that it is an extension of logical principles, maybe Occam's razor (itself discussed in TLP 3.328 and 5.47321).

However, the compatibility of options one and two – that is, the compatibility of law with logic – which seems to be supported in propositions 6.32 and 6.321, seems to be clearly contradicted by the following propositions:

> There is no possible way of making an inference from one situation to the existence of another, entirely different situation. There is no causal nexus to justify such an inference. We *cannot* infer the events of the future from those of the present. Superstition is nothing but belief in the causal nexus. (TLP 5.135–5.1361)

> The law of causality is not a law but the form of a law. 'Law of causality' – that is a general name. And just as in mechanics, for example, there are 'minimum-principles', such as the law of least action, so too in physics there are causal laws, laws of the causal form. (TLP 6.32–6.321)

The earlier propositions refer specifically to the question of the future and prediction, which are not issues mentioned in the later propositions. We could explain the above, by suggesting that the earlier propositions dispel the illusion that scientific laws state the future, and that the later statements refer to a correct form of scientific laws, as conditional statements of what will happen, given certain circumstances.

The question of the causal nexus cannot be explained in this way, though. If there are laws, there are causes. In any case, the dismissal of inference from one situation to another is more radical than a questioning of causal nexus. Even if we regard 'causal nexus' as a metaphysical assumption to be replaced by radical instrumentalism in scientific theory, we are still faced with the insistence on not being able to move from one situation to another. If Wittgenstein states that scientific laws are not laws of inference, this is surely too trivial. If Wittgenstein states that the links between events in scientific laws are unknown behind the phenomena and that the law is purely instrumental, he still does not need to deny an 'inferential link' between events. If he wishes to state that laws are fictional explanations of phenomena, there is nothing else in the *Tractatus* to suggest that this is the territory he wishes to enter, although once issues are raised in the way they are in the *Tractatus*, there is always the possible consequence of moving into such territory. Wittgenstein operates on the line between a metaphysical theory of logically founded laws of science – extreme nominalism, pragmatism and instrumentalism about the status of law.[34]

Philosophical Writing

TLP 5.1362 suggests a Kantian reason for limiting the scope of what Kant would call 'understanding':

> The freedom of the will consists in the impossibility of knowing actions that still lie in the future. We could know them only if causality were an *inner* necessity like that of logical inference. – The connexion between knowledge and what is known is that of logical necessity.

The limitations on causality, that it cannot have necessity or logical force, is what allows free will, because we cannot know how the future is determined.[35] This could mean that we are constitutively incapable of knowing, or that the future is undetermined.

Constantly in the *Tractatus*, Wittgenstein is ambiguous between a Kantian argument about the limits of our experience and an argument about reality in itself. The argument for free will clearly parallels Kant's argument that free will is established in showing the limits of phenomenal experience, and law-bound nature. According to Kant, there must be a noumenal self beyond phenomena, as there is for everything else. As the noumenal self must be beyond the causal laws of phenomenal nature, there must be a transcendental self and it must have free will.[36] Wittgenstein seems undecided between a Kantian transcendental argument and radical instrumentalism with regard to the status of law. Neither option is explicitly present, but these are the possible interpretations of Wittgenstein's remarks about law and free will. The Kantian transcendental, in Wittgenstein, is what can be shown but not said in questions of free will, ethics, the structure of the world and facts, language and propositions. Once we get beyond understanding, what refers to possible experience in Kant, we have dialectic: a form of philosophical writing exposing its own limits, and contradictions. This is what we have in the *Tractatus* in an unstated form. The ambivalences and contradictions, as with regard to the status of law, are a dialectical examination of the limits and contradictions of transcendental arguments. Wittgenstein distinguishes between the 'theoretical' reason of metaphysical contradictions and the 'practical' reason of propositions of TLP 6.41 to 6.4312. In contrast to Kant, this is not an absolute distinction as shown in proposition 6.41, for example, which must cover the theoretical and the practical:

> The sense of the world must lie outside the world. In the world everything is as it is, and everything happens as it does happen: *in* it no value exists – and if it did exist, it would

have no value. If there is any value that does have value, it must lie outside the whole sphere of what happens and is the case. For all that happens and is the case is accidental. What makes it accidental cannot lie *within* the world, since if it did it would itself be accidental. It must lie outside the world.

The mention of values provides a bridge between 'theoretical' and 'practical' concerns as Wittgenstein moves on to a discussion of ethics, aesthetics, religion and free will as outside the world, and moves towards subjective experience. It must also be the case that Wittgenstein is including the sense of law, logic, language, objects and so on, in the values that can only be expressed outside the world.[37] The distinction between the realm of theoretical dialectical illusion and the value of subjectivity is itself regulated by the impossibility of stating such transcendental claims without contradiction (the contradiction being that what is in the world can only be expressed as a whole outside the world, which is where there are properly values of what is not in the world). In Kant, of course, the absolute distinctions are not finally made as Kant has to keep repeating the distinctions and finding a way of grounding them without violating them – the whole of the *3rd Critique* is devoted to this task. A necessarily contradictory and dialectical task.

This establishes the case for seeing in Wittgenstein a journey from the Kant to Kierkegaard.[38] The Kantian starting point is the understanding and the attempt to limit reason to what is compatible with understanding, in which 'dialectic' marks the frontier of understanding in reason going beyond the limits of understanding. The Kierkegaardian conclusion is that dialectic is the condition of philosophy, in which philosophy is constituted by the incompatibility with subjective experience and abstract truth, although both are necessary for there to be communicable experience or comprehensible truth:

> Subjective reflection turns inward toward subjectivity and in this inward deepening will be of the truth, and in such a way that, just as in the preceding, when objectivity was advanced, subjectivity vanished, here subjectivity as such becomes the final factor and objectivity the vanishing. Here it is not forgotten, even for a single moment, that the subject is existing, and that existing is a becoming, and that truth as the identity of thought and being is therefore a chimera of abstraction and truly only a longing of creation, not because truth is not an identity, but because the knower is an existing person, and thus truth cannot be an identity for him as long as he exists. If this is not held fast, then with the aid of speculative thought we promptly enter into the fantastical *I=I* that recent speculative thought certainly has used but without explaining how a particular individual relates himself to it, and, good Lord, of course no human being is more than a particular individual.[39]

A full exploration of the journey from Kant to Kierkegaard would require discussion of Hegel, Fichte, Schelling and the Jena Romantics (Freidrich Schlegel, Karl Solger and Ludwig Tieck in particular).[40] Reasons of space preclude such a discussion in this paper, and in relation to Wittgenstein it is not strictly necessary as there is strong evidence that Wittgenstein knew his Kant and Kierkegaard,[41] but little biographically, or from hints in his writings, that he had studied these intervening philosophers and aesthetic thinkers. It is important to establish that there is a Kierkegaardian response to Kant implicit in the *Tractatus*, and that the study of Kierkegaard on truth, subjectivity, communication, system and religion is an appropriate response to reading the *Tractatus*. The consistent inner paradoxes of the *Tractatus* and declarations about its status as a ladder to be thrown away are Kierkegaardian. From Kierkegaard's point of view, philosophical writing and objective communication are necessarily prone to dialectic illusion. The illusions are worked through using paradox to expose their limitations and lead the way to the experience, which must be at the limit of writing and concepts:

> Before proceeding to the story of Abraham, I shall summon a pair of poetic individualities. With the power of dialectics, I shall hold them at the apex, and by disciplining them with despair, I may prevent them from standing still so that in their anxiety they may possibly be able to bring something or other to light.[42]

The *Tractatus* is a dialectic investigation of the impossibility of metaphysics. What the metaphysical investigations show goes beyond the purging of metaphysics. The statement of metaphysical theses is shown to be dialectic (that is, illusory), but only on condition that all abstract statements are dialectic and that all statements about abstract laws or objects are dialectical. This leads to the implication that philosophy is dialectical, but also to the claim, direct or indirect, that language is both correctly formed by nature and is dialectic. That seems like a perverse claim if we consider Wittgenstein's various claims that language is in order on its own, and that what can be said can be said simply. The claims about simple saying themselves contradict the role of saying in the *Tractatus*, which is the assertion of truth or falsity for what is shown in a proposition. If there are no statements about general objects, or abstractions of any kind, then language cannot be in order since the possibility of language is the possibility of abstraction. Every word or sentence in language is an abstraction, or universalization, of some kind. The *Tractatus* quite consistently contains an element of mystical retreat from language, which is in no way an irrelevant addition to the logical discussion as Russell, Hacker, and the other systematic readers claim.[43] The 'mystical' sections suggest that the *Tractatus* is a prelude to silence, modelled on Kierkegaard, or is a preface for a Kierkegaardian exploration of subjectivity.

Wittgenstein himself allowed the Russell type of response by failing to articulate clearly everything that is demonstrably implicit within the *Tractatus*, and being extraordinarily eager to allow the Logical Positivist and Russellian appropriation of the *Tractatus*. The latter is most starkly illustrated in the translation of '*Sachverhalten*' as 'atomic facts' in the Ogden translation (with the assistance of Frank Ramsey[44]) authorized by Wittgenstein, the first publication of the *Tractatus* in book form. 'States of affairs' in Pears and McGuinness is obviously preferable and less loaded.[45]

Wittgenstein himself could not find a clear way of articulating the relation between 'mysticism' and 'logic', allowing his own misreading and mistranslation. He was all too willing to allow the 'Logical Positivist' reading which allowed the *Tractatus* to be read as a manifesto for the reduction of philosophy to the clarification of logic and scientific statements reducible to observation statements. This is part of the history of the *Tractatus*; it has internal justification in many propositions and is not just a wrong reading. Nevertheless, it is a reading that ignores the context of such propositions. The propositions of that context belong with the 'positivist' propositions, which must be seen as what they are. They are so reductive that they expose the contradictions of a philosophy of logical and observational reduction: such a philosophy cannot state anything outside of the propositions of logical and observation; and the stating of such a philosophy contradicts itself.[46] The investigation of contradiction and transcendence in the *Tractatus* is that of the necessity of contradiction in positing a metaphysical view, as well as the inevitability of positing a metaphysical view when discussing language, objects, logic and so on, as these abstractions transcend the showing of their existence and become metaphysical. These Tractarian contradictions do not only show the limitations of the Logical Positivist project, with which it was associated for a while,[47] but of any attempt at a philosophical reduction to logic, concepts, semantic conditions or experience. It can most productively be seen as the continuation of reason's self-critique in Kant, and in later philosophical work on the limits and contradictions of pure reason.

Notes

* Thanks are due to Barry Allen for his private commentary on my paper; and to participants at a seminar at the Middle Eastern Technical University, Department of Philosophy, where I presented this paper, particularly Ahmet İnam, Geoff Bowe and John Bolander.

1 Immanuel Kant, *Critique of Pure Reason*, trans. P. Guyer and A. Wood (Cambridge: Cambridge University Press, 1997), pp. 386–87 [A297/B354–A298/B355].

2 There is a review of literature on Kant and Wittgenstein in Engel's 'Wittgenstein and Kant', *Philosophy and Phenomenological Research*, **39**, 1970, pp. 483–513. Engel criticizes earlier failures to perceive the importance of Kant to Wittgenstein and sketches out some parallels, with regard to morality, dialectic and transcendental arguments.

3 See Ludwig Wittgenstein, *Tractatus Logico-Philosophicus*, trans. C.K. Ogden (London: Routledge, 1922; German text with facing English text), 6.13. Hereafter referred to as 'TLP' in the text.

4 In this case, the Ogden (and Ramsey) translation is surely preferable: 'There is indeed the inexpressible. This *shows* itself; it is mystical'. The monistic sounding German which suggests something mystical is replaced in Pears and McGuinness by pluralistic English emphasizing things. The sense of a mystical space on the edge of language is replaced by things which are somehow mystical and unrepresentable, which makes much less sense in the context of the *Tractatus*.

5 In both translations of the *Tractatus* the second and third sentences are translated in the same way. This seems surprising as '*es sich verhält*' is translated in the plural, when surely it should be 'how it stands', not 'how things stand'. There is some similarity with TLP 6.522 (see n. 4 above), where it is just Pears and McGuinness who translate the 'Es' of the mystical as plural things. In both cases there seems to be some resistance to referring to a

general 'it' as an origin, rather than a particular empirical thing. This represents an irrepressible wish to create a more positivistic and empiricist Wittgenstein, which Wittgenstein of course collaborated in through his approval of the Ogden (and Ramsey) translation.

6 This adds to the evidence that Pears and McGuinness, Ogden (and Ramsey) prefer an empiricist and positivistic approach to the *Tractatus*, which emphasizes the truth-conditions of propositions, rather than the nature of the proposition itself and the transcendental possibilities of representation. See nn. 4 and 5 above.

7 Both the *Tractatus* translations insist on translating '*nämlich*' as 'truth'. While not an outrageous mistranslation, this insistence is interesting. There is an implicit anxiety for the translators with regard to the way in which Wittgenstein grounds what the conception of the world is saying. The transcendental implication that our world is constituted for us by language is an uncomfortable one, if Wittgenstein is treated as only concerned with militantly reductionist logicism and empiricism, encumbered by a few extraneous mystical remarks.

8 Erik Stenius suggests a solution to the complications of 'saying and showing' in *Wittgenstein's Tractatus: A Critical Exposition of the Main Lines of Thought*, X.2 (Bristol: Thoemmes Press, 1996), pp. 178–82. This relies on distinguishing between 'internal showing' of the logical form of elements and 'external showing' of the form of a state of affairs. However, this does not resolve the difficulties of why Wittgenstein varies between showing as all of language and showing as a component. Stenius gives a weak account of how there could be space for saying, by creating a more limited form of 'internal showing', but this confirms the problem rather than solving it. Now showing is determined as both external and internal, with a lack of a general account of 'showing' or 'saying'.

9 The Ogden (and Ramsey) translation is more literal for the second sentence: 'This connexion of the elements of the picture is called its structure, and the possibility of this structure is called the form of representation of the picture'. Pears and McGuinness impose the idea of a hypothetical means of expression in Wittgenstein's sentence. See Ludwig Wittgenstein, *Tractatus Logico-Philosophicus*, trans. B.F. McGuinness and D.F. Pears (London: Routledge, 1961).

10 This is a constant theme in Frege, in particular: see Gottlob Frege, 'Function and Concept', 'On Sense and Meaning', 'On Concept and Object', 'Logical Investigations', in *Collected Papers On Mathematics, Logic, and Philosophy*, ed. B. McGuinness, trans. M. Black *et al.* (Oxford: Blackwell, 1984), pp. 137–56, 157–77, 182–94 and 351–406 respectively.

11 Bertrand Russell, 'On Denoting', *Mind*, **14**, 1905, pp. 479–93 is the familiar reference here. It should also be noted that this issue, and others Wittgenstein considers in the *Tractatus*, are already present in Russell's *The Philosophy of Leibniz* (London: Routledge, 1992) first published in 1900. A discussion of the limits of the subject-predicate judgement and its limits can be found in Chapter 2, Section 11, pp. 16–23 of this work.

12 See Peter Geach, 'Saying and Showing in Frege and Wittgenstein', in J. Hintikka (ed.), *Essays on Wittgenstein in Honour of G.H. von Wright, Acta Philosophica Fennica* (Amsterdam: North Holland, 1976), pp. 54–70.

13 See Cora Diamond, 'Does Bismarck Have a Beetle in his Box? The Private Language Argument in the *Tractatus*', in A. Crary and R. Read (eds), *The New Wittgenstein* (London: Routledge, 2000), pp. 262–92.
14 Some discussion of the 'non-realist' nature of the *Tractatus* can be found in Roy Lemoine's *The Anagogic Theory of Wittgenstein's 'Tractatus'* (The Hague: Mouton, 1975), in particular, p. 39.
15 A very brief but penetrating discussion about states of affairs and the world appears in Alex Blum, 'Discussion: *Tractatus* 2.063', *Philosophical Investigations*, **12** (4), 1989, pp. 325–26.
16 Part 1, § 13 (Immanuel Kant, *Prolegomena to any Future Metaphysics*, trans. J.W. Beck (Indianapolis: Library of Liberal Arts, 1950), p. 33).
17 For example, Marie McGinn's 'Between Metaphysics and Nonsense: Elucidation in Wittgenstein's *Tractatus*', *Philosophical Quarterly*, **49**, pp. 491–513.
18 Kant, *Critique of Pure Reason*, *op. cit.*, n. 1, p. 394 [A310/B366–67].
19 Wittgenstein, *Tractatus*, *op. cit.*, n. 9, p. 3. This is anticipated by Hegel in *Hegel's Science of Logic*, trans. A.W. Miller (London: Allen & Unwin, 1969), Vol. 1, Bk 1, Sect. 1, Ch. 2, p. 134:

> With respect to the form of the *limitation* and the *ought*, two prejudices can be criticised in more detail. First of all, great stress is laid on the limitations of thought, of reason, and so on, and it is asserted that the limitation *cannot* be transcended. To make such an assertion is to be unaware that the very fact that something is determined as a limitation implies that the limitation is already transcended.

20 Ludwig Wittgenstein, *Philosophical Investigations*, ed. R. Rhees and G.E.M. Anscombe, trans. G.E.M. Anscombe, 2nd edn (Oxford: Blackwell, 1958).
21 Cora Diamond, *The Realistic Spirit: Wittgenstein, Philosophy and the Mind*, (Cambridge, MA: MIT Press, 1991).
22 James Conant, 'Throwing Away the Top of the Ladder', *Yale Review*, **79**, 1991, pp. 328–64.
23 Most commentators are systematic. The more notable include: G.E.M. Anscombe, *An Introduction to Wittgenstein's Tractatus* (London: Hutchinson, 1959); Max Black, *A Companion to Wittgenstein's Tractatus* (Ithaca, NY: Cornell University Press, 1964); Peter Carruthers, *Tractarian Semantics*, (Oxford: Blackwell, 1989); papers collected in Irving M. Copi and Robert W. Beard (eds), *Essays on Wittgenstein's Tractatus*, 2nd edn (New York: Hafner Press, 1973), in particular Gustav Bergmann, 'The Glory and the Misery of Ludwig Wittgenstein' and Wilfred Sellars, 'Naming and Saying'; P.M.S. Hacker, *Insight and Illusion: Themes in the Philosophy of Wittgenstein*, rev. edn (Oxford: Oxford University Press, 1989) reprinted by Thoemmes Press; idem, 'Was he trying to whistle it?', in Crary and Read, *The New Wittgenstein*, *op. cit.*, n. 13, pp. 353–88.
24 Diamond, *The Realistic Spirit*, *op. cit.*, n. 21, ch. 6.
25 See James Conant, 'Elucidation and Nonsense in Frege and Early Wittgenstein', in Crary and Read, *The New Wittgenstein*, *op. cit.*, n. 13, pp. 174–217.
26 Diamond, *The Realistic Spirit*, *op. cit.*, n. 21, p. 194.
27 A very sketchy reference is made to Wittgenstein's philosophical style in Cora Diamond's 'Ethics, Imagination and the *Tractatus*', in Crary and Read, *The New Wittgenstein*, *op. cit.*, n. 13, pp. 149–73.

28 Lynette Reid, 'Wittgenstein's Ladder: The *Tractatus* and Nonsense', *Philosophical Investigations*, **21** (2), 1998, pp. 97–151.
29 John Lippitt and Daniel Hutto, 'Making Sense of Nonsense: Kierkegaard and Wittgenstein', *Proceedings of the Aristotelian Society*, **XCVIII** (3), 1998, pp. 263–86.
30 Frege, 'On Sense and Meaning', in *Collected Papers, op. cit.*, n. 10, pp. 157–77).
31 This should certainly be compared with Brouwer's remarks about language, truth and art in sections V, VI and VII of 'Life, Art and Mysticism' (Chapter 1 of this volume).
32 On Tractarian ethics, and its relation with Kant, see Diamond's 'Ethics, Imagination and the *Tractatus*', *op. cit.*, n. 27.
33 *1st Critique*: B17–18 in Kant, *Critique of Pure Reason, op. cit.*, n. 1.
34 The relation between metaphysical law, possible worlds, logic and science is explored by Russell in *The Philosophy of Leibniz, op. cit.*, n. 11, which deserves to be studied in the context of the *Tractatus*. For example, section 29 (pp. 66–69) is very relevant to Wittgenstein's conception of logical space. In general, Russell's book on Leibniz can be seen as anticipating the *Tractatus* in the diagnosis of metaphysical illusion based on the incorrect understanding of propositions. One way of looking at the *Tractatus* is as a critical discussion of the metaphysical point of view which appears in Leibniz, the illusions inherent in this and the attempt to overcome illusion.
35 Wittgenstein's interest in free will outside causality is shared by Brouwer throughout 'Life, Art, and Mysticism' (Chapter 1 of this volume), but is stated in particular in Section II.
36 *2nd Critique*, Preface, in Immanuel Kant, *Critique of Practical Reason*, trans. L.W. Beck (New York: Macmillan/Library of Liberal Arts, 1993), pp. 3–14.
37 The separation of ethics from the world and from causality is suggested by Søren Kierkegaard in *Concluding Unscientific Postscript to Philosophical Fragments*, trans. E.V. Hong and E.H. Hong (Princeton, NJ: Princeton University Press, 1992), Pt 2, Sect. 2, Ch. 2: 'Insofar as the categories of cause and effect sometimes also take on a kind of semblance of guilt and punishment, this is only because the observer does not relate himself world-historically, is not able completely to strip away the ethical that is in him' (ibid, p. 155).
38 Stenius has some useful comments about Wittgenstein and Kant in *Wittgenstein's Tractatus*, XI, *op. cit.*, n. 8, pp. 214–26. However, Stenius remains at a very general level of comparison. He incorrectly assumes that Wittgenstein only knew Kant through Schopenhauer, and is generally too inclined to take Wittgenstein's claims to lack of knowledge and interest in the history of philosophy at face value. He certainly ignores Wittgenstein's interest in Kierkegaard and is lacking in the philosophical inclination to see that dialectical problems govern the whole of the *Tractatus*. Wittgenstein's comments about nonsense and throwing the ladder away are banalized into a commitment to reaching the end of philosophy.
39 Kierkegaard, *Concluding Unscientific Postscript, op. cit.*, n. 37, Vol. 1, pp. 196–97.
40 Kierkegaard's views on Hegel are expressed in various texts but particularly in *Concluding Unscientific Postscript, op. cit.*, n. 37. The relation of Kant with Fichte and the Jena Romantics is discussed by Kierkegaard in *The Concept of Irony with continual reference to Socrates*, trans. H.V. Hong and E.H. Hong (Princeton, NJ: Princeton University Press, 1989); his views on Schelling can be found in 'Notes of Schelling's Berlin Lectures' in the same volume.

41 This is confirmed in M.O'C. Drury's 'Some Notes on Conversations with Wittgenstein' and 'Conversations with Wittgenstein' in R. Rees (ed.), *Recollections of Wittgenstein*, (Oxford: Oxford University Press, 1984), pp. 76–171); also in Ray Monk's biography, *Ludwig Wittgenstein: The Duty of Genius*, (London: Random House, 1990).

42 Søren Kierkegaard, *Fear and Trembling/Repetition*, trans. H.V. Hong and E.H. Hong (Princeton, NJ: Princeton University Press, 1983), p. 88.

43 See Russell's 'Introduction' to the *Tractatus*, in Wittgenstein, *Tractatus*, *op. cit.*, n. 9, pp. ix–xxii); Hacker's *Insight and Illusion*, *op. cit.*, n. 23, p. 2.

44 See Monk, *The Duty of Genius* (*op. cit.*, n. 41, pp. 205–7) and Ogden's own acknowledgement in his prefatory 'Note' to the 1922 translation, *op. cit.*, n. 4.

45 A dissenting view and a fervent defence of Ogden and Ramsey can be found in: John O. Nelson, 'Is the Pears–McGuinness Translation of the *Tractatus* Really Superior to Ogden's and Ramsey's?', *Philosophical Investigations*, **22** (2), pp. 165–75.

46 This is why it is a mistake to claim that we can straightforwardly move from metaphysical to non-metaphysical philosophy on the basis of the *Tractatus*, as Hacker claims in *Insight and Illusion*, *op. cit.*, n. 23, p. 26.

47 See A.J. Ayer's Preface to *Language, Truth and Logic* (Harmondsworth: Penguin, 1971), p. 41.

- Not clear how the discussion bears on the question it was meant to address. (?)
- Probs with the trad. appr. are not recognized. (?)
- Nonsense is equated with contradictions.
- TLP as a Kierkegaardian dialectical investigation of the impossibility of metaph. (209, 210)

Chapter 10

Wittgenstein's Onto-Logic

Barry Allen

A *misunderstanding* makes it look to us as if a proposition *did* something queer.[1]

'To say of what is that it is, and of what is not that it is not, is true.'[2] Aristotle's statement of this classical formula seems unobjectionable, even platitudinous, yet it is not metaphysically innocent. It says truth is a sameness, an assimilation of *what is* (*ontos*) and *what is said* (*logos*). Moreover, this sameness is asymmetrical. Usually sameness is symmetrical: if *x* is the same as *y* (the same colour, say), *y* is the same as *x*. But if *x* is true to *y*, *y* is not true to *x*. As Aristotle explains elsewhere, a true proposition about a man 'is in no way the cause of the being of the man, but the fact of the man's being does seem somehow to be the cause of the truth of the proposition.'[3] It is up to us to 'tell it like it is', not up to beings to be as we say.

Between Parmenides and Aristotle, Greek philosophy elaborated this idea of truth – as receptive openness, passively contemplative transparency. Need, body, the contingencies of space, time, and circumstances withdraw, and things show themselves as they are, according to their own self-identical being. I call this idea of truth (and its problems) *onto-logic*, because it wants to ground the *logical* possibility of truth in language on the *ontic* possibility of the entity whose *being* (existence and identity) makes a discourse true.[4]

Wittgenstein's *Tractatus Logico-Philosophicus* unknowingly takes on the problems (and assumptions) of onto-logic, rediscovering its oldest principles. His later work is an equally unconscious deconstruction of the onto-logic he unknowingly recapitulated. The difference between early and late in Wittgenstein's work is therefore inscribed in two registers: his text, and his (unconscious) intertext, where it is an unknowing moment in the deconstruction of Western metaphysics.[5]

How Do Propositions Do It?

Two things can be the same or different colour only if both are coloured, and in that (second-order) respect *the same.* Things of same or different mass, duration or extension must one and all be massive, temporal, spatial and, in those ways, *the same*. Everything determinable in respect of same or different must be comparable, commensurable and thus, at a deeper level, *the same*. So, too, for the onto-logical sameness of truth. The mere possibility of truth-value – that what we say should be *the same* as what there is – requires that language and being are somehow *the same*.

Aristotle invented the wonderful idea of *form* (*eidos, morphē*) as the sameness that assimilates *nous* and *phusis*, mind and nature, body and soul, ensuring the possibility of science and truth. He defines *phusis*, nature, as 'shape (*morphē*) or form (*eidos*) according to *logos*.'[6] Whatever is present in nature thus accords with our *logos*, and vice versa, making Aristotle's definition of nature an intertextual commentary on a pregnant fragment of Parmenides: *it is the same thing to think and to be*.[7] There is no difference between *can be thought* (*said*) and *can be*; logic, *logos*, is the common measure.

Logos and *phusis*, language and nature, are thus assimilated, made one and the same through a concept of 'form'. The same thing happens to the soul, which Aristotle says 'is in a way all beings'. 'In a way,' because when one perceives a stone 'it is not the stone that is in the soul, but its form (*eidos*)'.[8] The soul is, in a way, *the same* as all beings, because it can assimilate any form, and form is something beings have to have. The *logos*-ordering of the soul and the *logos*-ordering of nature are the same. It is as Heraclitus said: one *logos* rules over all.[9] Nothing less ensures an onto-logical truth.

Dissatisfied with the epistemological weakness of Aristotle's idea (how do we know that the forms of our understanding are the veritable forms of reality and not mere appearance?), Kant reinterprets form as an ideational structure we impose on something unknowably different, abandoning metaphysical knowledge of true being for empirical prediction and control. His so-called Highest Principle of Synthetic Judgement – 'the a priori conditions of a possible experience in general are at the same time conditions of the possibility of objects of experience' – is a Cartesian reading of the Parmenidean text.[10] It says that the *forms* of our perception and understanding are the same as those of *phusis*, or phenomenal nature. To be (objective) is to be perceptible.

Dissatisfied with the idealistic entanglements that haunt Kant's text, Russell, Wittgenstein and the Logical Positivists took their linguistic turn, and interpreted the objectivity of knowledge as a matter of logic and syntax, or what they called correct *logical form*. Wittgenstein explicitly thinks logical form onto-logically, as the assimilating sameness of language and reality. 'There must be something identical in a picture and what it depicts to enable the one to be a picture of the other at all' (TLP 2.161); and it is logical form: 'What any picture ... must have in common with reality in order to be able to depict it – correctly or incorrectly – in any way at all is logical form, which is the form of reality' (TLP 2.18).[11] One *logos* rules alike in language, mind, and reality. 'A thought', he says, 'contains the possibility of the situation of which it is the thought. What is thinkable is possible too' (TLP 3.02). The *logos* of thought is the *logos* of being, as Parmenides said; and 'form' is a sort of sameness that assimilates mind and nature, as Aristotle explained.

This logical form is absolute and subsistent. It is not the mere artefact of an optional meta-language, not relative to a choice of languages or logical theories, and explains more than the inference relations among propositions. It must account for the primordial possibility of an onto-logical truth, where what is said is exactly the same as a fact in the logical space of the world. What the *Tractatus* calls logical form is the respect in which arbitrary symbols and actual situations can be the same in the way that makes one true. Such form is common to all (first-order) respects of sameness and difference (space, time, colour and so on), and is in that way a second-order sameness, a sameness among respects of sameness. Differences of colour, time, extension, and so on are so many determinations of the one 'logical form' common to language and reality. Being, meaning the totality not just of actual, but *possible*, existence, coincides with the possible determinations of 'logical form.'

The first-order respects of sameness and difference (for example, colour, shape, duration) are what make empirical (and especially 'elementary') propositions definitely true or false. What Wittgenstein calls the objects (*Gegendstände*) of any definitely true-or-false proposition are not everyday things but logically simple moments of sameness or difference, – for example, one colour, extension, duration, or pitch.

In Kant, an 'object' is an object of experience, meaning an object of representation and judgement, whose 'objectivity' is formally postulated by an empirical concept. Such objects are the familiar themes of everyday experience: causes, effects, substances in space and time. Familiar though they are, for Kant these 'objects' are not the true forms of being, which are unknown to us. Kant speaks of 'representations' (*Vorstellungen*), but he is the first major thinker to set aside the old idea of perception and concept as *resemblance.* Kantian representations do not *resemble* anything. They do not reflect, mirror-like, the identity or truth of a being.

For Wittgenstein, that is exactly what a true proposition does – the queer thing it accomplishes. He goes behind Kant, and back to Aristotle (if not to Heraclitus), to assimilate language and reality in the service of onto-logical truth. Although he calls propositions pictures (*Bilder*), he thinks of them less as representations (*Vorstellungen*) than as copies, models or projections – symbols whose very presence is a mimesis of being. Underlying his view of what a proposition does – 'reaching right out to reality' – is a view of what objects are and how simple signs manage to name them. According to the theory of the *Tractatus*, this feat of naming requires that sign and object be somehow the same. But that is strange! Same in what way? 'What does the logical identity of sign and thing signified really consist in? This question is (*once again*) a main aspect of the whole philosophical problem' (NB 7).[12]

Like Aristotle (whom he proudly claimed never to have read), Wittgenstein offers a two-track explanation, relying on a concept of form to locate one and the same *logos* in language and reality. Objects and names are logical mirror-images of each other. Consider first objects, the ground-floor constituents of any real situation. As I read him, Wittgenstein identifies these objects with simple – that is, not further determinable – determinations of logical form. An object, in this sense, might be a colour or location, a duration or a mass, while the situation of a colour and a mass at a location for an interval would be a complex of objects, or what he calls a 'state of affairs' (*Sachverhalt*). In confirmation I cite the only two passages where Wittgenstein says something informative about the identity of his 'objects':

> Space, time, and colour (being coloured) are forms of objects. (TLP 2.0251)
>
> A spatial object must be situated in infinite space. (A spatial point is an argument place.) A speck in the visual field, though it need not be red, must have some colour: it is, so to speak, surrounded by colour-space. Notes must have *some* pitch, objects of the sense of touch *some* degree of hardness, and so on. (TLP 2.0131)

He is speaking here of *forms of object*, and doing so in terms appropriate to simple first-order determinations of the second-order sameness he calls logical form.

How many such forms are there? Why does Wittgenstein merely mention 'space, time, colour', as it were in passing, and not be more precise? Because it is, he thinks, no part of logic to say what there is. It belongs to the '*application* of logic' to decide what elementary propositions there are, and hence what forms and what objects make them true; and 'what belongs to its application logic cannot anticipate' (TLP 5.557). Unlike Aristotle and Kant, Wittgenstein has no Doctrine of Categories or Transcendental Deduction. 'Logical forms are *without* number. Hence there are no pre-eminent numbers in logic, and hence no philosophical monism or dualism, etc.' (TLP 4.128).

This conception of the object – a logically simple moment of sameness or difference – is matched by a parallel account of how names signify simple determinations of first-order difference. Aristotle used the idea of form to assimilate nature and the soul, and I have suggested that Wittgenstein does something similar. But he pressed the problem further than Aristotle, who was content to endow the soul with wonderful powers. Like an engineer, Wittgenstein wonders how something could *do* what Aristotle says the soul does – be the same as what it perceives – or how propositions do the queer thing they do, reaching right out to reality.

In pursuit of an answer Wittgenstein discovers something Saussure also did at about the same time: a principle of original difference.[13] For Saussure, it explains the relationship between a sign (signifier) and the concept it signifies. It explains how a conceptual field can be articulated by the material of language, primarily speech. In Wittgenstein, original difference has a more metaphysical application. It explains not the relation between word and concept, but the relation between language and reality.

'If two objects have the same logical form, the only distinction between them, apart from their external properties, is that they are different' (TLP 2.0233). For instance, if in a situation with two objects of the same form (colours, say), we abstract from their contingent relationship (red beside blue), the only thing that makes them *two* are their differences – all of them, from every other simple determination of that form. This differential constitution makes an object the opposite of a self-identical atom. Objects are not primarily identical to themselves and then contingently different from each another. There is nothing *to* an 'object' (or the situations they comprise) apart from systematic difference. Hence, 'if objects are given, then at the same time we are given *all* objects' (TLP 5.524).

That is why Wittgenstein emphasizes systematic difference as essential both to signs and to things. Of things, he says: 'Each thing is, as it were, in a space of possible states of affairs. This space I can imagine empty' – for example, nothing actually coloured – 'but I cannot imagine the thing [the colour] without the space' (TLP 2.013) – that is, the differences.[14] As for signs, 'Where one can construct symbols according to a system, logical importance attaches to the system and not the individual symbols' (TLP 5.555) – since what the symbols are, what they name or mean, depends on a *logos* of difference visible only in the whole, the system. What is 'logical' about signs is how their use is differentially related in a whole language; and once signs *are* so differentiated they cannot help referring to (mimicking) an object (that is, a possible difference in reality). There are no 'mistakes' in logic – nothing with the wrong name, or unnamed – because:

> ... the system of rules which determine a calculus thereby determines the meaning [*Bedeutung*] of its symbols. More correctly expressed: The form and the syntactic rules are equivalent. If I change the rules – seemingly supplementing or completing them – I thereby change the form, the meaning.[15]

Original difference is supposed to illuminate how a proposition can be true to the world, which it does by explaining how names (in propositional context) represent the ground-floor objects of any situation in the world. Wittgenstein exploits the same possibility that attracted Saussure – namely, that units as different materially as a sound and a concept can share the same differential syntax. The distinctness of the objects that make propositions definitely true-or-false depends not on what each is for itself (self-identity), but how all differ among themselves, and this 'form,' this *logos* of difference, can be *reproduced*, *the same*, in the differential syntax of a sign.

Any sign (token) is a bundle of properties, some of which function symbolically and some not. For example, the colour of a printed predicate does not function symbolically and is arbitrary to the meaning. Denotative signs can be defined as those for which every property functioning symbolically in the symbolizing material is arbitrary in that way. All the symbolizing properties of 'wet' are arbitrary in relation to its meaning; and while 'polysyllabic' is polysyllabic and in that respect non-arbitrarily related to its meaning, nothing in the logic of denotation requires this sameness, which does not function in the determination of meaning.

Things are otherwise for copies, reproductions or models. Their form, their organization, their syntax (how they are put together) is not *arbitrary* in relation to their meaning; it is the same. A copy is not a 'conventional' (that is, arbitrary) representation, but actually represents what it is about or says (to be), as a sample of colour or cloth really has the qualities it refers to.[16] A copy shows the very thing and, if true, is the very same as what it means. And that, of course, is exactly what Wittgenstein says a proposition does. 'One name stands for one thing, another for another, and they are combined with one another. In this way the whole group – like a *tableau vivant* – presents a state of affairs' (TLP 4.0311).

How does it do it, this queer, this astonishing, accomplishment? Any logically simple sign will have qualities that are arbitrary in relation to its meaning (the object it names), but these do not function symbolically in the sign; and the properties that *do* so function are so far from arbitrary that they are actually the very same as the meaning, this sameness functioning symbolically in the sign as the means by which reference is determined. What properties are these? Wittgenstein tells us when he says:

> In order to recognize a symbol by its sign we must observe how it is used with a sense. A sign does not determine a logical form unless it is taken together with its logico-syntactical employment'. (TLP 3.326–27)

That differential syntax is not arbitrary in relation to the referent: it is the same, the same *logos*, the same form of difference. Name and object are one and the same determination of one and the same logical form, systematically differing (from other names and objects), and in this difference finding their identity as elements, whether of language or reality. That is why, 'although there is something arbitrary in our notations, *this* much is not arbitrary – that *when* we have determined one thing arbitrarily, something else is necessarily the case' (TLP 3.342). What is not arbitrary is the *logos* of differentiation exemplified by any sign. There is nothing arbitrary about entire languages, which could not be different in a single one of their logical aspects.[17]

Because the very 'being' of what there is to get right or be true to is originally differential, and shares this origin with differences in language, it is possible for a simple sign (in the context of a proposition) to replicate, reproduce and thus *exemplify* (and in that way refer to and name) an 'element' of being. Name and object can be the same, and a propositional configuration of names definitely the same or different from the facts in logical space. A sign can exemplify a colour *difference* regardless of how it looks, since even a 'look' – that is, a moment of perceptual consciousness – 'means' or 'shows' a colour (say), not by its immediate phenomenal presence, but by its transcendental-psychological syntax.

A proposition is comparable to the world in the same way that a copy is comparable to an original, a sample to a suit, or a ruler to the extension it measures. It 'has the same kind of relation to reality that a measuring rod has to an object. This is not a simile; the measuring rod is an example of the relation'.[18] A proposition has a truth-value, or a truth-determining comparability to the world, because the names that comprise it are little samples of the objects whose combination coincides with the facts in logical space. '*That* is how a picture is connected to reality ... It is laid against reality like a yardstick' (TLP 2.1511–12). If the world is the same as the model, if it is no larger or smaller than the yardstick, then what it shows veritably is, and what is said by its construction (thinking) is true.

The *Aufhebung* of Logic

Not just this solution, but the entire fabric of assumptions that create the onto-logical problem it addresses, receive far-reaching criticism in Wittgenstein's later work. Consider his later meditations on standards, like a standard metre or standard sepia. Many of these passages can be read as parables about what the *Tractatus* called objects.[19] Most of their formal features reappear: one cannot describe a standard (with the predicate it standardly determines), but can only give it a name (TLP 3.221) – that is, a logico-syntactical employment (TLP 3.327). The standard is what has to exist if predication (colour, length and so on) has a determinate sense (TLP 3.23). It is simple (TLP 2.02), – that is, exactly one determination of exactly one logical form (one length, one colour). And since the language game we play with a standard simply takes its existence for granted, it may even seem 'unalterable and subsistent' (TLP 2.027).[20]

One interest of these later reflections is the dawning awareness they show of how the fact of practice is bedrock. If we have the language game, the standard and its practice, we don't need the supplement of substance. We don't need 'objects' to make our statements true. They don't add anything, or make a difference, to the language game. What *has* to exist, if there is true-or-false, is the practice, which is contingent, *and nothing else besides.* 'Like everything metaphysical,' Wittgenstein later writes, 'the harmony between thought and reality is to be found in the grammar of the language.' 'What looks as if it *had* to exist belongs to the language. It is a paradigm in our language game; something with which comparison is made.'[21]

In his first work Wittgenstein mistook the onto-logical intertext of Western metaphysics for an intuition philosophy has to respect. Consider something he once said 'forces itself upon us *over and over again*' – the necessity of positing 'something Simple, Indivisible, an element of being, in brief a thing'.

> We feel that the WORLD must consist of elements ... that the world must be what it is, it must be determinate.... It seems as if to deny things were as much as to say: The world

> can be as it were indeterminate, in something like the sense in which our knowledge is uncertain and indeterminate. The world has a fixed [*feste*] structure. (NB 62)

The later work shakes off this intuition. The 'logical possibility' of truth originates in contingent practice, rests (if you want to call it 'resting') on contingent practice. There are no objects, logical forms, elementary propositions or atomic names. Indeed, logic itself *seems to be abolished.*

In his use of the word *Logik* Wittgenstein is closer to Hegel and Heraclitus than Tarski or Quine. 'Logic is not a doctrine [*Lehre*], but a mirror-image of the world. Logic is transcendental' (TLP 6.13). It belongs at once to language and reality, transcending and assimilating their difference. In a text redolent of Heraclitus Wittgenstein calls this deep order *das Logische*:

> What is logical in what is signified [*das Logische des Bezeichneten*] should be completely determined just by what is logical in the sign and the method of symbolizing [*das Logische des Zeichens und der Bezeichnungsweise*]. One might say: sign and method of symbolizing together must be logically identical with what is signified. (NB 19)

A proposition *says* 'what is', but it *shows* what *can be* (TLP 4.022), which requires that the syntactic possibility of its formation be the same as the *ontological* possibility of a situation which, if actual, makes the combination true. The order of language and the order of being are the same. 'My *whole* task,' Wittgenstein once wrote, 'consists in explaining the nature of the proposition. That is to say, in giving the nature of all facts, whose picture the proposition *is.* In giving the nature of all being [*Sein*]. (And here being does not mean existing [*existieren*] – in that case it would be nonsensical)' (NB 39). What is or is not an existing being logic cannot say, and doesn't need to ('it would be nonsensical'). Logic is concerned with the being of whatever there is, which for Wittgenstein means what is logical about a being. This 'logic' is *onto*-logic, dedicated to the assimilation of language and being, and is *ontological* in Heidegger's proper sense, meaning concerned with *being* in its *difference* from entities, or what is.[22]

'Logic must take care of itself.' That is how Wittgenstein tried to think its sublime transcendence, first in a notebook entry (22 August 1914): 'Logic must take care of itself. If syntactical rules for functions can be set up *at all*, then the whole theory of things, properties, etc., is superfluous.... Once more: logic must take care of itself. A *possible* sign must also be capable of signifying.' The next entry, the next week: 'It must in a certain sense be impossible for us to go wrong in logic. This is already partly expressed by saying: Logic must take care of itself. This is an extremely profound and important insight' (NB 2). Then in the *Tractatus*: 'Logic must take care of itself. If a sign is possible, then it is also capable of signifying.... In a certain sense, we cannot make mistakes in logic' (TLP 5.473).

Logic must take care of itself says two things. First, that logical language and possible being are the same, and that therefore what seems logical is logical; hence there are no 'purely logical' mistakes. 'We can think nothing unlogical [*nichts Unlogisches*], since, if we could, we would have to think unlogically' (TLP 3.03). If that is impossible, it must be that thinking is operating logically with signs, combining them in a logical way, which means an onto-logical way, to mimic or model a situation in the logical space of the world.

'If sign and thing signified were *not* identical in respect of their total logical contents,' Wittgenstein writes in his notebook, 'then there would have to be something still more fundamental than logic' (NB 4). It could not take care of itself. Logic would not be logic if the 'possibility' of truth depended on something 'still more fundamental than logic'. Yet there is something more fundamental, although 'fundament' is not the right picture, since this 'something more' is an abyss of contingency. What has to exist if there is truth in language is nothing sublime, transcendent or unsayable beyond language. All that really *has* to exist is a language-game, a usage, a contingent practice. Truth-value has nothing to do with onto-logical sameness, and everything to do with mutual expectations in the use of language. Agreement among people, not in judgements (where they differ all the time), but in their habits, games and culture, ensures a certain constancy in the use of language and makes truth a banal accommodation not to the forms of reality, but to the speech of others.

With sardonic mock horror, Wittgenstein says his argument 'seems to abolish logic' [*scheint die Logik aufzuheben*].[23] Of course, nothing could 'abolish logic,' which is absurd, but the argument does abolish (or render pointless) the *onto*-logic he once mistook for the purest logic. 'The more closely we examine actual language, the sharper becomes the conflict between it and our demand. (For the crystalline purity of logic was certainly not the *result* of an investigation; it was a demand.)'[24] Logic is more banal than sublime, more like natural history than transcendental mechanics. There is no more force to the necessity of a 'necessary truth' than the normativity enforced in symbolizing practice. The impression of necessity for, say, $4 + 1 = 5$ merely registers the (normative) decision to use this proposition as a criterion of 'knowing how to add' – that is, for being able to play the language game.[25] And if the language-game with the standard sepia did not exist, there would be no residual determination *in reality* that would be inaccessible to a thus impoverished symbolism. If we start making rulers out of dough, we are false to nothing except our own earlier practice. 'The way the world is' has nothing to do with it.

Scheint die Logik aufzuheben. Logic may not be abolished, but the onto-logic of the *Tractatus* is, precisely, *aufgehoben.* It is relieved of its reason, deconstructed. Its onto-logical postulates – object, situation, logical form – lose their rationale, become fifth wheels, don't make a difference. They *seemed* indispensable, but that turns out to be a mistake, depending, as it does, on the onto-logical misunderstanding that in being true or false a proposition *does* something *queer.*[26]

Notes

1 Ludwig Wittgenstein, *Philosophical Investigations*, ed. R. Rhees and G.E.M. Anscombe, trans. G.E.M. Anscombe, 2nd edn (Oxford: Blackwell, 1958), para. 93.
2 Aristotle, *Metaphysics* 1011b.
3 Aristotle, *Categories* 14b.
4 On onto-logic, see my *Truth in Philosophy* (Cambridge, MA: Harvard University Press, 1993), especially ch. 1.
5 On intertextuality, see Julia Kristeva, *Desire in Language* (New York: Columbia University Press, 1980), pp. 65–69; and Michael Riffeterre, 'The Intertextual Unconscious', *Critical Inquiry*, **13**, 1987. Borrowing from Russian theorist Mikhail Bakhtin, Kristeva introduced this concept into semiotics, where it was widely taken up in the study of poetic language. Intertextuality is the transposition of one or more system of signs into another, accompanied by a new articulation of the combined elements. This is a normal feature of language-use, which cannot be aseptically protected from the force of signs whose home base and order may be distant and heterogeneous to the context in which they appear. For an application of the idea to philosophical texts, see my 'The Historical Discourse of Philosophy', *Reconstructing Philosophy: New Essays in Metaphilosophy, Canadian Journal of Philosophy*, supp., **19**, pp. 127–58. Further, 'deconstruction' seems the right image for what Wittgenstein describes when he writes: 'Where does our investigation gets its importance from, since it seems only to destroy everything interesting, that is, all that is great and important? (As it were all the buildings, leaving behind only bits of stone and rubble.) What we are destroying is nothing but houses of cards and we are clearing up the ground of language on which they stand' (*Philosophical Investigations*, *op. cit.*, n. 1, § 118).
6 Aristotle, *Physics* 193a.
7 Parmenides, fragment 3, in K. Freeman (ed.), *Ancilla to the Pre-Socratic Philosophers* (Cambridge, MA: Harvard University Press, 1948), p. 42.
8 Aristotle, *De Anima* 431b.
9 'All things come to pass in accordance with this logos'; 'It is wise, listening not to me but to the *logos*, to agree that all things are one' (Charles H. Kahn, *The Art and Thought of Heraclitus* (Cambridge: Cambridge University Press, 1979), an edition of the fragments with translation and commentary, I/1; XXXVI/50 (Kahn/Diels)).
10 Immanuel Kant, *Critique of Pure Reason*, trans. N. Kemp Smith (London: Macmillan, 1933), A111; cf. A158/B197.
11 Ludwig Wittgenstein, *Tractatus Logico-Philosophicus*, 2nd edn, trans. B.F. McGuinness and D.F. Pears (Oxford: Blackwell, 1961), 2.161, 2.18. Referred to hereafter as TLP in the text.
12 Ludwig Wittgenstein, *Notebooks, 1914–1916*, 2nd edn, ed. G.H. von Wright and G.E.M. Anscombe, trans. G.E.M. Anscombe, (Chicago: University of Chicago Press, 1979), p. 3. Hereafter referred to as NB in the text.
13 See my 'Difference Unlimited', in G.B. Madison (ed.), *Working Through Derrida* (Evanston: Northwestern University Press, 1993).
14 This seems to be a point Rudolf Carnap took from his study of the *Tractatus*: 'The reality of anything is nothing else than the possibility of its being placed in a certain system' (*Philosophy and Logical Syntax* (London: Kegan Paul, 1935), p. 20).
15 Ludwig Wittgenstein, *Philosophical Remarks*, trans. R. Hargreaves and R. White (Chicago: University of Chicago Press, 1975), p. 178.

16 On Exemplification as a mode of reference, see Nelson Goodman, 'Routes of Reference', in his *Of Mind and Other Matters* (Cambridge, MA: Harvard University Press, 1984); and *Languages of Art*, 2nd edn (Indianapolis: Hackett, 1976), chs 2–5. For him, a sample refers not to a property, but to a label which then denotes items in a resemblance class that includes the sample. Hence, for him, an exemplification is not the same as its referent in the referred-to respect, since that referent is not a property but a label, or predicate.

17 Cf. G.W.F. Leibniz: 'I know that the Scholastics and everyone else are given to saying that the significations of words are arbitrary (*ex instituto*), and it is true they are not settled by natural necessity; but they are settled by reasons' (*New Essays on Human Understanding*, trans. P. Remnant and J. Bennett (Cambridge: Cambridge University Press, 1981), p. 278).

18 Ludwig Wittgenstein, *Wittgenstein's Lectures, Cambridge 1930–1932, From the Notes of John King and Desmond Lee*, ed. D. Lee (Chicago: University of Chicago Press, 1982), p. 6. In another place he refers to *Tractatus* 2.1511: 'I have elsewhere said that a proposition "reaches up to reality," and by this I meant that the forms of the entities are contained in the form of the proposition which is about these entities' ('Remarks on Logical Form', in I. Copi and R.W. Beard (eds), *Essays on Wittgenstein's Tractatus*, (New York: Macmillan, 1966), p. 36.

19 The most important text for my point is *Philosophical Investigations*, *op. cit.*, n. 1, § 50; also § 242 and p. 225. For comparable or complementary passages see Ludwig Wittgenstein, *Philosophical Grammar*, trans. A. Kenny (Berkeley: University of California Press, 1974), pp. 142–43; idem, *Remarks on the Foundations of Mathematics*, rev. edn, trans. G.E.M. Anscombe (Cambridge, MA: MIT Press, 1978), pp. 40–41, 167–68, 382; idem, *Wittgenstein's Lectures on the Foundations of Mathematics, Cambridge 1939*, ed. C. Diamond (Hassocks, Sussex: Harvester, 1976), pp. 55, 73; and idem, *Remarks on the Philosophy of Psychology*, Vol. 1, ed. G.E.M Anscombe and G.H. von Wright, trans. G.E.M. Anscombe (Chicago: University of Chicago Press, 1980), §§ 445, 632. These are not all the places where Wittgenstein says something about measuring or standards, although they are places where attention is discreetly drawn to the formal similarity between standards and what the *Tractatus* called 'objects'.

20 'It … serves as a standard; this makes it aloof and non-temporal.' (Wittgenstein, *Wittgenstein's Lectures on the Foundations of Mathematics*, *op. cit.*, n. 19, p. 73.)

21 Wittgenstein, *Philosophical Grammar*, *op. cit.*, n. 19, p. 112; and *Philosophical Investigations*, *op. cit.*, n. 1, § 50.

22 On Heidegger's idea of ontology see Allen, *Truth in Philosophy*, *op. cit.*, n. 4, ch. 5.

23 Wittgenstein, *Philosophical Investigations*, *op. cit.*, n. 1, § 242.

24 Ibid., § 107.

25 See Wittgenstein, *Remarks on the Foundations of Mathematics*, *op. cit.*, n. 19, pp. 319–20. I am not saying that this normativity can be seen 'from nowhere'. It can be articulated only from the inside and for insiders, as Robert Brandom insists in *Making it Explicit* (Cambridge, MA: Harvard University Press, 1994).

26 For further discussion of my interpretation, see the chapter on Wittgenstein in Allen, *Truth in Philosophy*, *op. cit.*, n. 4.

- correct leg. form (218)

Bibliography

Allen, Barry (1993), 'Difference Unlimited', in G.B. Madison (ed.), *Working Through Derrida*, Evanston: Northwestern University Press, pp. 5–27.

Allen, Barry (1993), *Truth in Philosophy*, Cambridge, MA: Harvard University Press.

Allen, Barry (1994), 'The Historical Discourse of Philosophy', *Reconstructing Philosophy: New Essays in Metaphilosophy, Canadian Journal of Philosophy*, supp., **19**, pp. 127–58.

Anscombe, G.E.M. (1959 and 1963), *An Introduction to Wittgenstein's Tractatus*, 1st and 2nd edns, London: Hutchinson and Philadelphia: University of Pennsylvania Press.

Ayer, A.J. (1971), *Language, Truth and Logic*, Harmondsworth: Penguin.

Baker, G.P. and Hacker, P.M.S. (1985), *Wittgenstein: Rules, Grammar and Necessity*, Oxford: Blackwell.

Barrett, Cyril (1991), *Wittgenstein on Ethics and Religious Belief*, Oxford: Blackwell.

Bell, David (1996), 'Solipsism and Subjectivity', *European Journal of Philosophy*, **4**, pp. 155–74.

Black, Max (1964), *A Companion to Wittgenstein's Tractatus*, Ithaca, NY: Cornell University Press.

Blum, Alex (1989), 'Discussion: Tractatus 2.063', *Philosophical Investigations*, **12**(4), pp. 325–26.

Bouwsma, O.K. (1965), *Philosophical Essays*, Lincoln: University of Nebraska Press.

Brandom, Robert B. (1994), *Making it Explicit*, Cambridge, MA: Harvard University Press.

Brockhaus, Richard R. (1991), *Pulling up the Ladder: The Metaphysical Roots of Wittgenstein's Tractatus Logico-Philosophicus*, La Salle: Open Court.

Carnap, Rudolf (1935), *Philosophy and Logical Syntax*, London: Kegan Paul.

Carruthers, Peter (1989), *Tractarian Semantics*, Oxford: Blackwell.

Cavell, Stanley (1976), *Must We Mean What We Say?: A Book of Essays*, Oxford: Oxford University Press.

Cavell, Stanley (1979), *The Claim of Reason: Wittgenstein, Skepticism, Morality and Tragedy*, Oxford: Oxford University Press.

Cavell, Stanley (1984), 'Existentialism and Analytic Philosophy', in *Themes Out of School*, San Francisco: North Point Press.

Cavell, Stanley (1989), 'An Interview with Stanley Cavell by James Conant', in R. Fleming and M. Payne (eds), *The Senses of Stanley Cavell*, Bucknell Review, Lewisburg, PA: Bucknell University Press, pp. 21–72.

Cavell, Stanley (1989), *This New Yet Unapproachable America: Lectures after Emerson after Wittgenstein*, Albuquerque: Living Batch Press.

Cavell, Stanley (1990), *Conditions Handsome and Unhandsome: The Constitution of Emerson Perfectionism*, Chicago: University of Chicago Press.

Cavell, Stanley (1996), 'The *Investigations*' Everyday Aesthetics of Itself', in S. Mulhall (ed.), *The Stanley Cavell Reader*, Oxford: Blackwell.

Conant, James (1989), 'Must We Show What We Cannot Say,' in R. Fleming and M. Payne (eds), *The Senses of Stanley Cavell*, Bucknell Review, Lewisburg, PA: Bucknell University Press, pp. 242–83.

Conant, James (1991), 'The Search for Logically Alien Thought: Descartes, Kant, Frege and the *Tractatus*', *Philosophical Topics*, **20** (1), pp. 101–80.

Conant, James (1991), 'Throwing Away the Top of the Ladder', *Yale Review*, **79**, pp. 328–64.

Conant, James (1992), 'Kierkegaard, Wittgenstein and Nonsense', in T. Cohen, P. Guyer and H. Putnam (eds), *Pursuits of Reason*, Lubbock: Texas Technical University Press, pp. 195–224.

Conant, James (1995), 'Putting Two and Two Together: Kierkegaard, Wittgenstein and the Point of View for Their Work as Authors', in T. Tessin and M. von der Ruhr (eds), *Philosophy and the Grammar of Religious Belief*, London: St Martin's Press and Basingstoke: Macmillan, pp. 248–331.

Conant, James (1996), 'Putting Two and Two Together: Kierkegaard, Wittgenstein and Nonsense', in D.Z. Phillips (ed.), *The Grammar of Religious Belief*, New York: St Martin's Press.

Conant, James (1998), 'Wittgenstein on Meaning and Use', *Philosophical Investigations*, **21**(3), pp. 222–50.

Conant, James (2000), 'Elucidation and Nonsense in Frege and Early Wittgenstein', in A. Crary and R. Read (eds), *The New Wittgenstein*, London: Routledge, pp. 174–217.

Conant, James (2001), 'A Prolegomenon to the Reading of Later Wittgenstein', in C. Mouffe and L. Nagl (eds), *Deconstruction and Pragmatism*, Frankfurt am Main: Peter Lang Press.

Conant, James (2001), 'Carnap and Early Wittgenstein on Logical Syntax', in T. McCarthy (ed.), *Wittgenstein in America*, Oxford: Oxford University Press.

Conant, James (2001), 'On Going the Bloody Hard Way in Philosophy', in J. Whittaker (ed.), *The Possibilities of Sense*, New York: Macmillan.

Conant, James (2001), 'The Method of the *Tractatus*', in E.H. Reck (ed.), *From Frege to Wittgenstein: Perspectives on Early Analytic Philosophy*, Oxford: Oxford University Press.

Conant, James (2002), 'What "Ethics" in the *Tractatus* is *Not*', in D.Z. Phillips (ed.), *Wittgenstein on Ethics and Religion*, New York: St Martin's Press.

Copi, I.M. and Beard, R.W., (eds) (1966 and 1973), *Essays on Wittgenstein's 'Tractatus'*, 1st edn, New York: Macmillan; 2nd edn, New York: Hafner Press.

Coudert, A.P. *et al.* (eds) (1998), *Leibniz, Mysticism and Religion*, Dordrecht: Kluwer.

Crary, Alice and Read, Rupert (eds) (2000), *The New Wittgenstein*, London: Routledge.

Diamond, Cora (1991), 'Ethics, Imagination and the Method of Wittgenstein's *Tractatus*', in R. Heinrich and H. Vetter (eds), *Bilder der Philosophie*, *Wiener Reihe* 5, Vienna: Oldenbourg, pp. 55–90.

Diamond, Cora (1991), *The Realistic Spirit: Wittgenstein, Philosophy and the Mind*, Cambridge, MA: MIT Press.

Diamond, Cora (2000), 'Does Bismarck Have a Beetle in his Box? The Private Language Argument in the *Tractatus*', in A. Crary and R. Read (eds), *The New Wittgenstein*, London: Routledge.

Diamond, Cora (2000), 'Ethics, Imagination and the *Tractatus*', in A. Crary and R. Read (eds), *The New Wittgenstein*, London: Routledge, pp. 149–73.

Edwards, J. (1982), *Ethics Without Philosophy: Wittgenstein and the Moral Life*, Tampa: University Presses of Florida.

Engel, S. Morris (1970), 'Wittgenstein and Kant', *Philosophy and Phenomenological Research*, **39**, pp. 483–513.

Engelmann, Paul (ed.) (1968), *Letters from Ludwig Wittgenstein: With a Memoir*, New York: Horizon Press.

Finch, Henry (1977), *Wittgenstein: The Later Philosophy: An Exposition of the 'Philosophical Investigations'*, New Jersey: Humanities Press.

Flew, Anthony (1951), *Logic and Language*, Oxford: Blackwell.

Floyd, Juliet (1998), 'The Uncaptive Eye: Solipsism in Wittgenstein's *Tractatus*', in L.S. Rouner (ed.), *Loneliness*, Boston University Studies in Philosophy and Religion, Notre Dame: University of Notre Dame Press, pp. 79–108.

Freeman, K. (ed.) (1948), *Ancilla to the Pre-Socratic Philosophers*, Cambridge, MA:Harvard University Press.

Frege, Gottlob (1980), *The Foundations of Arithmetic*, trans. J.L. Austin, Evanston, Ill.: Northwestern University Press.

Frege, Gottlob (1984), *Collected Papers on Mathematics, Logic, and Philosophy*, ed. B. McGuinness, trans. M. Black *et al.*, Oxford: Blackwell.

Frege, Gottlob (1997), *The Frege Reader*, ed. M. Beaney, Oxford: Blackwell.

Geach, P.T. (1976), 'Saying and Showing in Frege and Wittgenstein', in J. Hintikka (ed.), *Essays on Wittgenstein in Honour of G.H. von Wright*, *Acta Philosophica Fennica* 28, Amsterdam: North Holland, pp. 54–70.

Geach, P.T. (1991), *Philosophical Encounters*, ed. H. Lewis, Dordrecht: Kluwer.

Geach, P.T. and Black, M. (eds) (1984), *Translations from the Philosophical Writings of Gottlob Frege*, Oxford: Blackwell.

Glock, Hans-Johann (1996), *A Wittgenstein Dictionary*, Oxford: Blackwell.

Goodman, Nelson (1976), *Languages of Art*, 2nd edn, Indianapolis: Hackett.

Goodman, Nelson (1984), *Of Mind and Other Matters*, Cambridge, MA: Harvard University Press.

Gustafsson, Martin (unpublished manuscript), 'Nonsense and Philosophical Method'.

Hacker, P.M.S. (1986), *Insight and Illusion: Themes in the Philosophy of Wittgenstein*, Oxford: Clarendon Press; rev. edn, Oxford: Oxford University Press: 1989; reprinted, Bristol: Thoemmes, 1997.

Hacker, P.M.S. (1996), *Wittgenstein's Place in Twentieth-Century Analytic Philosophy*, Oxford: Blackwell.

Hacker, P.M.S. (1999), 'Naming, Thinking and Meaning in the *Tractatus*', *Philosophical Investigations*, **22** (2), pp. 119–35.

Hacker, P.M.S. (2000), 'Was He Trying to Whistle It?', in A. Crary and R. Read (eds), *The New Wittgenstein*, London: Routledge, pp. 353–88.

Hacker, P.M.S. (unpublished manuscript), 'When the Whistling Had to Stop'.

Hacker, P.M.S. and Baker, G. (1980), *An Analytical Commentary on Wittgenstein's Philosophical Investigations*, Vol. 1, Oxford: Blackwell.

Hacker, P.M.S. and Baker, G. (1990), *Wittgenstein: Meaning and Understanding*, Oxford: Blackwell.

Hegel, G.W.F. (1969), *Hegel's Science of Logic*, trans. A.W. Miller, London: Allen & Unwin.Hintikka, Jaakko (1958), 'On Wittgenstein's "Solipsism"', *Mind*, **67**.

Hutto, Daniel (1995), 'Consciousness Demystified: A Wittgensteinian Critique of Dennett's Project', *The Monist*, **78** (4), pp. 464–78

Hutto, Daniel (1996), 'Was the Later Wittgenstein a Transcendental Idealist?', in P. Coates and D. Hutto (eds), *Current Issues in Idealism*, Bristol: Thoemmes Press, pp. 121–53.

Hylton, Peter (1997), 'Functions, Operations and Sense in Wittgenstein's *Tractatus*', in W. Tait (ed.), *Early Analytic Philosophy*, Chicago: Open Court, pp. 91–105.

Jacquette, Dale (1990), 'Wittgenstein and the Colour Incompatibility Problem', *History of Philosophy Quarterly*, **7** (3), pp. 353–65.

James, William (1961), *The Varieties of Religious Experience: a Study in Human Nature*, New York: Collier Books.

John of the Cross (1935), 'Points of Love', in *The Complete Works of Saint John of the Cross*, Vol. III, ed. E.A. Peers, trans. P.S. de Santa Teresa, London: Burns, Oates and Washbourne Ltd.

Johnston, Peter (1989), *Wittgenstein and Moral Philosophy*, London: Routledge.

Kahn, Charles H. (1979), *The Art and Thought of Heraclitus*, Cambridge: Cambridge University Press.

Kant, Immanuel (1933), *Critique of Pure Reason*, trans. N. Kemp Smith, London: Macmillan.

Kant, Immanuel (1950), *Prolegomena to any Future Metaphysics*, trans. J.W. Beck, Indianapolis: Library of Liberal Arts.

Kant, Immanuel (1952), *The Critique of Judgement*, trans. J.C. Meredith, Oxford: Oxford University Press.

Kant, Immanuel (1993), *Critique of Practical Reason*, trans. L.W. Beck, New York: Macmillan/Library of Liberal Arts.

Kant, Immanuel (1997), *Critique of Pure Reason*, trans. P. Guyer and A.W. Wood, Cambridge: Cambridge University Press.

Kenny, Anthony (1973), *Wittgenstein*, Harmondsworth: Penguin.

Kenny, Anthony (1982), 'Wittgenstein's on the Nature of Philosophy', in B. McGuinness (ed.), *Wittgenstein and His Times*, Oxford: Blackwell.

Kierkegaard, Søren A. (1962), *The Point of View for My Work as an Author*, trans. W. Lowrie, New York: Harper & Row.

Kierkegaard, Søren A. (1983), *Fear and Trembling/Repetition*, trans. H.V. Hong and E.H. Hong, Princeton, NJ: Princeton University Press.

Kierkegaard, Søren A. (1989), *The Concept of Irony with continual reference to Socrates*, with 'Notes of Schelling's Berlin Lectures', trans. H.V. Hong and E.H. Hong, Princeton, NJ: Princeton University Press.

Kierkegaard, Søren A. (1992), *Concluding Unscientific Postscript to Philosophical Fragments* (2 vols), trans. E.V. Hong and E.H. Hong, Princeton, NJ: Princeton University Press.

Koethe, John (1996), *The Continuity of Wittgenstein's Thought*, Ithaca, NY: Cornell University Press.

Koethe, John (unpublished manuscript), 'Making Sense of the *Tractatus*'.

Kremer, Michael (1997), 'Contextualism and Holism in the Early Wittgenstein: From *Prototractatus* to *Tractatus*', *Philosophical Topics*, **25**, pp. 87–120.

Kremer, Michael (2001), 'The Purpose of *Tractarian* Nonsense', *Noûs*, **35**, pp. 39–73.

Kremer, Michael (2002), 'Mathematics and Meaning in the *Tractatus*', *Philosophical Investigations*, **20** (3), pp. 272–303.

Kristeva, Julia (1980), *Desire in Language*, New York: Columbia University Press.

Leibniz, G.W.F. (1948), *Textes Inédits*, Vol. 1, ed. G. Grua, Paris: Presses Universitaires de France.

Leibniz, G.W.F. (1981), *New Essays on Human Understanding*, trans. P. Remnant and J. Bennett, Cambridge: Cambridge University Press.

Leibniz, G.W.F. (1984), *Discours de Métaphysique et Correspondance avec Arnauld*, ed. G. LeRoy, Paris: Librairie Philosophique J. Vrin.

Leibniz, G.W.F. (1989), *Philosophical Essays*, ed. and trans. R. Ariew and D. Garber, Indianapolis and Cambridge: Hackett Publishing.

Lemoine, Roy E. (1975), *The Anagogic Theory of Wittgenstein's 'Tractatus'*, The Hague: Mouton.

Lewy, C. (1967), 'A Note on the Text of the *Tractatus*', *Mind*, **76**.

Lippitt, John and Hutto, Daniel (1998), 'Making Sense of Nonsense: Kierkegaard and Wittgenstein', *Proceedings of the Aristotelian Society*, **XCVIII** (3), pp. 263–86.

Luckhardt, C.G. (ed.) (1979), *Wittgenstein: Sources and Perspectives*, Ithaca, NY: Cornell University Press; reprinted, Bristol: Thoemmes Press, 1996.

Malcolm, Norman (1986), *Nothing Is Hidden: Wittgenstein's Criticism of his Early Thought*, Oxford: Blackwell.

Malcolm, Norman and Winch, Peter (1993), *Wittgenstein: A Religious Point of View?*, London: Routledge.

Mayer, Verena (1993), 'The Numbering System in the *Tractatus*', *Ratio*, **6** (2), pp. 108–20.

McDonough, Richard (1994), 'A Note on Frege's and Russell's Influence on Wittgenstein's *Tractatus*', *Russell*, **3** (14), pp. 39–46.

McGinn, Marie (1997), *Wittgenstein and the Philosophical Investigations*, London: Routledge.

McGinn, Marie (1999), 'Between Metaphysics and Nonsense: Elucidation in Wittgenstein's *Tractatus*', *The Philosophical Quarterly*, **49**, pp. 491–513.

McGuinness, Brian (1976), 'The Grundgedanke of the *Tractatus*', in *Understanding Wittgenstein*, Royal Institute of Philosophy Lectures, 1972/73, vol. 7, London: Macmillan.

McGuinness, Brian (1981), 'The So-called Realism of the *Tractatus*', in I. Block (ed.), *Perspectives on the Philosophy of Wittgenstein*, Oxford: Blackwell.

McGuinness, Brian (1988), *Wittgenstein: A Life; Young Ludwig 1889–1921*, Berkeley: University of California Press.

Monk, Ray (1990), *Ludwig Wittgenstein: The Duty of Genius*, New York: Free Press and London: Random House.

Mounce, H.O. (1997), 'Philosophy, Solipsism and Thought', *Philosophical Quarterly*, **47**, pp. 1–18.

Mulhall, Stephen (1994), *Stanley Cavell: Philosophy's Recounting of the Ordinary*, Oxford: Clarendon Press.

Nelson, John O. (1999), 'Is the Pears–McGuinness Translation of the *Tractatus* Really Superior to Ogden's and Ramsey's?', *Philosophical Investigations*, **22** (2), pp. 165–75.

Pears, D.F. (1987), *The False Prison: A Study of the Development of Wittgenstein's Philosophy*, vol. 1, Oxford: Clarendon Press.

Pears, D.F. (1996), 'The Originality of Wittgenstein's Investigation of Solipsism', *European Journal of Philosophy*, **4**, pp. 124–36.

Peterson, Donald (1990), *Wittgenstein's Early Philosophy: Three Sides of the Mirror*, Toronto: University of Toronto Press.

Reid, Lynette (1998), 'Wittgenstein's Ladder: The *Tractatus* and Nonsense', *Philosophical Investigations*, **21**(2), pp. 97–151.

Rhees, Rush (1967), 'A Symposium: Assessments of the Man and the Philosopher', in K.T. Fann (ed.), *Ludwig Wittgenstein*, New York: Delta Publishing.

Rhees, Rush (ed.) (1984), *Recollections of Wittgenstein*, Oxford: Oxford University Press.

Riffeterre, Michael (1987), 'The Intertextual Unconscious', *Critical Inquiry*, **13**.

Russell, Bertrand (1905), 'On Denoting', *Mind*, **14**, pp. 479–93; reprinted in R.C. Marsh (ed.), *Logic and Knowledge*, London: Unwin, 1956, pp. 41–56.

Russell, Bertrand (1918), 'The Philosophy of Logical Atomism'; reprinted in R.C. Marsh (ed.), *Logic and Knowledge*, London: Unwin, 1956, pp. 177–281.

Russell, Bertrand (1959), *My Philosophical Development*, London: Routledge.

Russell, Bertrand (1992), *The Philosophy of Leibniz*, London: Routledge.

Schopenhauer, Arthur (1969), *The World as Will and Representation*, Vol. 1, trans. E.F.J. Payne, New York: Dover.

Silesius, Angelus (1946), *Pélerin Chérubinique (Cherubinischer Wandersmann)*, ed. and trans. H. Plard, Paris: Aubier.

Stenius, Erik (1996), *Wittgenstein's Tractatus: A Critical Exposition of the Main Lines of Thought*, Oxford: Blackwell, 1960; reprinted, Bristol: Thoemmes Press, 1996.

Sullivan, Peter M. (1996), 'The "Truth" in Solipsism and Wittgenstein's Rejection of the A Priori', *European Journal of Philosophy*, **4**, pp. 195–219.

Teresa of Avila (1991), *The Life of Teresa of Jesus: The Autobiography of Teresa of Avila*, ed. and trans. E.A. Peers, New York: Image Books.

Thomas, Dylan (1982), *Poems*, ed. D. Jones, London: J.M. Dent.

Thomas, Emyr Vaughan (1999), 'From Detachment to Immersion: Wittgenstein and the "Problem of Life"', *Ratio*, **12**, pp. 195–209.

Thompson, Michael (1995), 'The Representation of Life', in R. Hursthouse, G. Lawrence and W. Quinn (eds), *Virtues and Reasons*, Oxford: Oxford University Press.

Tilghman, B.R. (1991), *Wittgenstein Ethics and Aesthetics: The View From Eternity*, Basingstoke: Macmillan.

Weil, Simone (1963), *Gravity and Grace*, London and New York: Routledge.

Weiner, Joan (1991), *Frege in Perspective*, Ithaca, NY: Cornell University Press.

Weininger, Otto (1906), *Sex and Character*, authorized translation from the 6th German edn, London: William Heinemann.

Williams, Meredith (1998), *Wittgenstein's Meaning and Mind*, London: Routledge.

Winch, Peter (ed.) (1969), *Studies in the Philosophy of Wittgenstein*, London: Routledge.

Winch, Peter (1972), *Ethics and Action*, London: Routledge.

Wittgenstein, Ludwig (1922), *Tractatus-Logico Philosophicus*, trans. C.K. Ogden, London: Routledge. (German text with facing English text.)

Wittgenstein, Ludwig (1929), 'A Lecture on Ethics'; reprinted in J. Klagge and A. Nordmann (eds), *Ludwig Wittgenstein: Philosophical Occasions 1912–1951*, Indianapolis: Hackett, 1993.

Wittgenstein, Ludwig (1931 and 1980), *Culture and Value*, ed. G.H. von Wright and H. Nyman, trans. P. Winch, Oxford: Blackwell.

Wittgenstein, Ludwig (1935) (unpublished), 'Lectures on Personal Experience', ed. C. Diamond.

Wittgenstein, Ludwig (1958), *Philosophical Investigations*, ed. R. Rhees and G.E.M. Anscombe, trans. G.E.M. Anscombe, 2nd edn, Oxford: Blackwell.

Wittgenstein, Ludwig (1961), *Notebooks, 1914–1916*, ed. G.H. von Wright and G.E.M. Anscombe, trans. G.E.M. Anscombe, Oxford: Blackwell; New York: Harper and Row, and Oxford: Blackwell, 1969; 2nd edn, Chicago: University of Chicago Press, 1979.

Wittgenstein, Ludwig (1961), *Tractatus Logico-Philosophicus*, trans. B.F. McGuinness and D.F. Pears, London: Routledge.

Wittgenstein, Ludwig (1967), *Zettel*, 2nd edn, ed. G.E.M. Anscombe and G.H. von Wright, trans. G.E.M. Anscombe, Oxford: Basil Blackwell.

Wittgenstein, Ludwig (1969), *The Blue and Brown Books*, 2nd edn, Oxford: Basil Blackwell.

Wittgenstein, Ludwig (1969), *On Certainty*, ed. G.E.M. Anscombe and G.H. von Wright, trans. D. Paul and G.E.M. Anscombe, Oxford: Blackwell.

Wittgenstein, Ludwig (1973), *Letters to C.K. Ogden*, ed. G.H. von Wright, Oxford: Blackwell.

Wittgenstein, Ludwig (1974), *Philosophical Grammar*, trans. A. Kenny, Berkeley: University of California Press.

Wittgenstein, Ludwig (1975), *Philosophical Remarks*, trans. R. Hargreaves and R. White, Chicago: University of Chicago Press.

Wittgenstein, Ludwig (1976), *Wittgenstein's Lectures on the Foundations of Mathematics, Cambridge 1939*, ed. C. Diamond, Hassocks, Sussex: Harvester; Chicago: University of Chicago Press, 1989.

Wittgenstein, Ludwig (1978), *Remarks on the Foundations of Mathematics*, rev. edn, trans. G.E.M. Anscombe, Cambridge, MA: MIT Press.
Wittgenstein, Ludwig (1980), *Remarks on the Philosophy of Psychology*, Vol. I, ed. G.E.M. Anscombe and G.H. von Wright, trans. G.E.M. Anscombe, Oxford: Blackwell and Chicago: University of Chicago Press.
Wittgenstein, Ludwig (1980), *Remarks on the Philosophy of Psychology*, Vol. II., ed. G.H. von Wright and P. Nyman, trans. C.G. Luckhardt and M.A.E. Aue, Oxford: Blackwell and Chicago: University of Chicago Press.
Wittgenstein, Ludwig (1980), *Wittgenstein's Lectures, Cambridge 1930–1932, From the Notes of John King and Desmond Lee*, ed. D. Lee, Oxford: Blackwell; Chicago: University of Chicago Press, 1982.
Wittgenstein, Ludwig (1982), *Last Writings on the Philosophy of Psychology*, Vol. I, ed. G.H. von Wright and H. Nyman, trans. C.G. Luckhardt and M.A.E. Aue, Oxford: Blackwell.
Wittgenstein, Ludwig (1984), *Tractatus Logico-Philosophicus*, Frankfurt: Suhrkamp.
Wittgenstein, Ludwig (1991), *Geheime Tagebücher 1914–1916*, ed. W. Baum, Vienna: Turia & Kant.
Wittgenstein, Ludwig (1993), *Philosophical Occasions 1912–1951*, ed. J. Klagge and A. Nordmann, Indianapolis: Hackett.
Wittgenstein, Ludwig (1997), *Denkbewegungen: Tagebücher 1930–1932, 1936–1937. Teil 1: Normalisierte Fassung*, ed. I. Somavilla, Innsbruck: Haymon Verlag.

Index

- p. 102: ¶ on the role of nonsense: illum. nons. leads to the adoption of the correct logical p. of view. (cf 110) 107, 115
  - No. The corr. p. of view is not expressible in props. It's embodied in the notation. The not. is justified by our recogn. of the collapse of TLP's props. on the basis of our ordinary log. capacity.
- Transcending TLP's props. Cf. 108.
- To possess concepts or a notation is to be able to make certain kinds of statements (& perhaps not feeling the temptation to make other statements to which a logically equivalent not. leads because of being less perspic.)
- Diamond: TLP use the not strictly corr. meth. Ref. cf. p. 147, note 35.
- 137: Acc. to therapy reading trans. per. & colour exclusion probl. can't be made any sense of.
- 137, 138: form's of life as transcendent basis of lang. (switch from one theory of lang. to another.)
- Add Hutto to constructivists. (140)
- Misleading pic of thinking as inner process PI § 316. (Mention as example in Ch. 1.)
- PI § 69 drawing a boundary for a spec. purpose.
- TLP's ethic is the ethics of clarity. The sol. to probs. lies in their disappearance.
- Quasi-truth (61)